£21·60.

This b
be re
belov
it ma
visit
Plea:
for
com;

Fines

Book

16 SEP.

Monographs in Developmental Biology

Vol. 11

Editor-in-Chief:
A. WOLSKY, New York, N. Y.

Co-Editors:
B. M. CARLSON, Ann Arbor, Mich.; P. S. CHEN, Zürich;
T. A. DETTLAFF, Moscow; G. REVERBERI, Palermo;
J. G. SCANDALIOS, Raleigh, N. C.

S. Karger · Basel · München · Paris · London · New York · Sydney

Histogenesis and Morphogenesis in Planarian Regeneration

ROSINE CHANDEBOIS

Laboratory of Animal Morphogenetics, University of Provence, Marseille

50 figures and 1 table, 1976

S. Karger · Basel · München · Paris · London · New York · Sydney

SPECIAL

Monographs in Developmental Biology

Vol. 7: Poglazov, B. F. (Moscow): Morphogenesis of T-Even Bacteriophages. VI + 105 p., 56 fig., 6 tab., 1973.
ISBN 3–8055–1645–2

Vol. 8: Myers, R. B. and Cantino, E. C. (East Lansing, Mich.): The Gamma Particle. A Study of Cell-Organelle Interaction in Development. X + 117 p., 43 fig., 12 tab., 1974.
ISBN 3–8055–1735–1

Vol. 9: McGeachie, J. K. (Nedlands, W. A.): Smooth Muscle Regeneration. IX + 90 p., 39 fig., 7 tab., 1975.
ISBN 3–8055–2058–1

Vol. 10: Baccetti, B. (Siena) and Afzelius, B. A. (Stockholm): The Biology of the Sperm Cell. VI + 254 p., 19 fig., 26 pl., 1 tab., 1976.
ISBN 3–8055–2204–5

Cataloging in Publication
Chandebois, Rosine
Histogenesis and morphogenesis in planarian regeneration by Rosine Chandebois. Basel, New York, Karger, 1976.
(Monographs in developmental biology; v. 11)
1. Turbellaria – growth and development 2. Turbellaria – cytology 3. Regeneration.
W1 M0567L v. 11/QX 350 C454h
ISBN 3–8055–2285–1

© Copyright 1976 by S. Karger AG, Basel (Switzerland), Arnold-Böcklin-Strasse 25
Printed in Switzerland by Merkur AG, Langenthal
ISBN 3–8055–2285–1

Contents

Preface

«Du choc des opinions
jaillit la vérité.»

'Truth springs from the clash of opinions.' This French saying is no-where more true than in science. Often real progress can only be made after established theories have been challenged. And is it not so that in biology, due to the very nature of living things, many 'theories' are hardly more than considered opinions?

The amazing regenerative powers of flatworms have always intrigued biologists because they so strikingly illustrate biological organization at the supracellular level, perhaps so far the most elusive of all biological phenomena. Planarian regeneration is a comparatively old subject, and perhaps for that reason a number of established opinions have accumulated and have finally found their way into treatises now considered classical. This, let us face it, can be a dangerous situation, because it may block the way towards significant new insight.

For a quarter of a century Prof. CHANDEBOIS has been working with the most varied types of planarians, both fresh-water and marine, using classical and modern approaches. She has gradually come to view their regenerative processes, and particularly the cellular phenomena underlying them, from a radically different angle than most of her predecessors and contemporaries do. In this book she aptly summarizes the evidence for her views, most of which was assembled by her and her students at Marseille. The confrontation with these unorthodox ideas has been a most refreshing experience, even (or perhaps particularly) for one who is not among the few real experts in this difficult area. The author's views simply cannot be ignored, and moreover deserve to be known outside the limited circle of those working on regeneration, in particular to those interested in the organism's reactions to injury or stress.

As so often happens, it is clear in retrospect that the author was not entirely alone with her ideas, for some of them may already be found in more or less rudimentary form in the publications of some of the most competent workers of a few decades ago. The contemporary literature also occasionally furnishes clues that point in the same direction.

Some of the author's notions are rather abstract, perhaps necessarily so in view of the complexity of the processes involved. Although I do not always entirely agree with what she says, I am nevertheless convinced that her book is an important contribution and that it will stimulate much new thinking and experimentation.

J. FABER
Hubrecht Laboratory, Utrecht

Acknowledgements

It is indeed a very great pleasure and honour for me to dedicate this book to my teacher, Prof. MARCEL ABELOOS, as a token of my gratitude. During the many long years that he directed my research on planarians, he instilled in me the necessity of reworking a theory each time a result could not be fitted within its confines. His example taught me the humility of a theoretician. When he has worked out his idea, he must not be content to repose upon it, or to fight to preserve it at all cost. He must offer it to the younger researchers for whom it will serve as a stepping stone allowing them to reach still higher in the unceasing search for truth.

It is natural that new ideas should be treated with mistrust and scepticism. This is a necessary precaution to restrain theorists with too much imagination; but it also presents a danger of sclerosis. Consequently my deep felt thanks go to Prof. ALEXANDER WOLSKY for his suggestion that I write this monograph on the regeneration of planarians. He offered me this possibility of regrouping results that were scattered in many short papers, of confronting them with current ideas, of rethinking them in the light of earlier works long since forgotten despite their importance, and of comparing them with information that we now possess for other animal groups. I would also like to thank Prof. WOLSKY for the interest with which he followed the developments of the manuscript, for all his advice and his friendly encouragements.

My sincerest gratitude goes to all those who helped me with the actual writing of my monograph: to Prof. JEAN BRACHET who offered the benefit of his great knowledge of nucleic acids and who willingly spent many hours reviewing the chapter on histogenesis; to Dr. JACOB FABER whose enthusiasm inspired me during many animated discussions on the problem of morphogenesis and whose competence in amphibian regeneration was of great help to me; and to Dr. BRUCE CARLSON who undertook to review the manuscript, and offered many useful suggestions.

I cannot forget all my associates who backed up my research, especially FRANCIS BECCHERINI, technician in electron microscopy and all the scientists in my laboratory whose work on regeneration enlightened my own research: ROSE MARIE COULOMB-GAY, FRED HOARAU, JEAN PIERRE CORNEC, MICHEL HIRN and MABROUK TURKI. My sincerest thanks and my friendship go to each of them.

I thank the publisher S. Karger AG for the great care with which they brought the reproduction of drawings and micrographs.

Introduction

The name planarian, rather than *Triclade,* has become widespread for designating freely moving flatworms that belong to the group Turbellaria and are characterized by ciliated epidermis and no anus. Their distinguishing feature is their digestive system, which consists of three blind-ending branches. These animals are quite different from other groups in the morphology of the adult and in their peculiar embryonic development which takes place within a cocoon from lose cellular material. The general organization of planarians and their classification is thoroughly described in classic textbooks of zoology. Thus we will simply supply a few diagrams (fig. 1) and refer the reader to the general treatises [DE BEAUCHAMP, 1961; HYMAN, 1951; KÜKENTHAL, 1928–1933; LAMEERE, 1932] for more detailed information.

Life for a planarian begins with a very unusual embryogenesis (fig. 2): the succession of epigenetic processes is completely different from those in other metazoans. The most recent descriptions of planarian development [LE MOIGNE, 1963; KOSCIELSKI, 1966] have not contradicted the strange observations recorded by the first researchers [HALLEZ, 1887; MATTIESEN, 1883]. The egg, distinguishable by the strong basophily of its cytoplasm, is enclosed within a chitinous cocoon along with a large number of 'vitelline' cells, which, according to PRENANT [1922] contain no yolk. When cleavage is about to begin, the 'vitelline' cells close to the egg elongate along the axis perpendicular to its surface and become rich in RNA. While the first blastomeres are dispersing, the transformed 'vitelline' cells unite, forming a syncytium characterized by its basophily and very small nuclei. Somewhat later several blastomeres migrate to the periphery of the syncytium, forming a single layer. The embryonic area thus comes to look like a blastula. A second migration of blastomeres towards the periphery sets up a temporary ectoderm. Meanwhile other cells are grouped in an excentric position and differentiate to form a temporary pharynx. Then the remaining exterior 'vitelline' cells are engulfed. In the meantime a pocket is hollowed out in the syncytium to receive these cells and tempo-

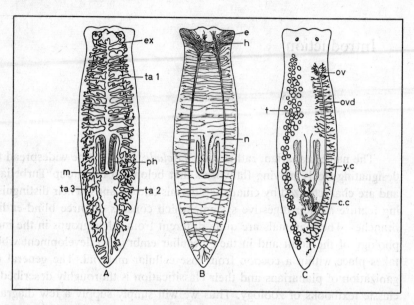

Fig. 1. Morphology of *Dendrocoelum lacteum*. *A* Digestive tract and excretory organs. *B* Nervous system. *C* Reproductive system. c. c = Copulatory complex; e = eyes; ex = excretory organs; h = head ganglia; m = mouth; n = nerve trunks; ov = ovary; ovd = oviduct; ph = pharynx; t = testes; ta 1–ta 3 = the three branches of the intestine; y. c = yolk cells. From LAMEERE.

rary endoderm appears on its walls. This is thus an exceptional process: the 'vitelline' syncytium is secondarily integrated with the embryo. It serves as a mesenchyma and contains free cells derived from the first blastomeres which migrated into the syncytium. This whole organization is temporary. A new pharynx appears in front of or behind the temporary pharynx which degenerates. There also appears a new epidermis. SKAER [1965] believes that this is due to a massive proliferation of cells coming from the mesenchyma and then added to the already existing cells. The endoderm is also replaced. As for the 'vitelline' syncytium [PRENANT, 1922], it liquefies and becomes more homogeneous; the nuclei break up and the whole syncytium is replaced by a parenchyma in which cell boundaries are everywhere easily distinguishable. All these formations, including the primordium of a nervous system, are produced by the same cell type: blastomeres scattered throughout the 'vitelline' syncytium. This is the start of a *unique* histogenetic system which will henceforth remain unchanged. From now on, only parenchymal cells will multiply and differentiate, in order to re-

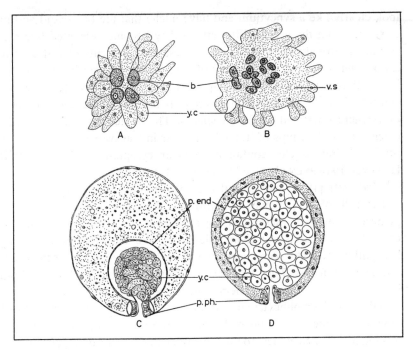

Fig. 2. Development of triclads. *A* Four-cell stage. *B* Blastomeres scattered in the vitelline syncytium at a later stage of cleavage. *C, D* Absorption of yolk cells. b = Blastomere; p. end = provisional endoderm; p. ph. = provisional pharynx; v. s = vitelline syncytium; y. c = yolk cells. A, B from LAMEERE; B, C from HYMAN.

place the cells in all the tissues. This situation is quite exceptional for coelomates. In other forms the cells of any of the embryonic layers can produce only those differentiated types which they normally provide. They cannot replace types normally produced by another germ layer. The peculiar exception which the planarians represent allows us to explain the persistence of totipotency in cells, a fact commonly accepted by embryologists.

In the adult, the parenchyma (fig. 6B) fills up all the spaces between the epidermis and the internal organs. There is no coelom and no blood circulation. PRENANT [1922] distinguished, in this particular tissue, a network comparable to connective tissue of other groups as well as free cells, mostly blood cells, that move about freely in it. The so-called fixed parenchyma is largely cellular; only the youngest, almost undifferentiated cells look clearly like a syncytium and have nuclei that are big, ovoid and hard to stain. When these cells differentiate they become elongated fibers with shrunken strongly stainable nuclei. These two elements of different ages form a network made up entirely of cells and almost no intercellular collagen. PEDERSEN's observations [1961] with electron microscope partly confirmed these results. In the fixed parenchyma the membranes are everywhere separated by 200 Å spaces. The nuclei are few. The cellular extensions – or groups of 'fibers' – appear in transverse section as a multitude of tiny vesicles containing a clear cytoplasm and mitochondria. However, PEDERSEN observed no syncytial structures.

PRENANT [1922] carried out the only existing detailed cytologic study of free cells of the parenchyma. Although there is no circulatory system in planarians, most of these cells are analogous to certain vertebrate blood cells. The nucleus is not clearly delimited and the cytoplasm is strongly basophilic. PRENANT recognized two main types, related to each other by intermediary forms. *Type I cells* (fig. 22A), whose basophilic cytoplasm is scanty and whose nuclear membrane is not clearly defined, correspond exactly to the lymphocytes of vertebrates. *Type II* is the equivalent of hemoblasts of the vertebrate embryo. PRENANT made the assumption that totipotent cells, to which for a long time the faculty of repairing all the tissues of the organism (especially those of the genital glands at the beginning of each period of sexual activity) has been attributed, belong to the type II cells. Although they are clearly the same type as blood cells of the vertebrates, PRENANT supposed that they have retained the properties of embryonic cells and are still capable of taking part in the formation and restoration of somatic and germinal tissues.

Planarians probably owe their exceptional regenerative faculty to the totipotent character of certain of the parenchymal cells. In the case of most species studied, any part of the body can be reconstituted when it has been cut off, be the cut transversal, longitudinal or oblique. Thus, the little fragments that represent less than one thirtieth of the body can reconstruct a complete and normal worm. The regenerative faculty, contrary to general belief, is not a leftover from the capacity of embryonic regulation. In fact, regenerative faculty appears fairly late in the course of development, when a certain degree of organization is reached. BARDEEN [1902] showed that embryos can reconstitute an amputated cephalic region only when their nervous system is already formed at the time of the operation. Moreover, the regeneration processes are very different from embryonic regulation, which is simply a spatial redistribution of constituents in the

germ: they are more complicated in their apparent modalities and determination.

If one removes a cephalic region of a planarian by cutting transversally right behind the eyes, the wound normally heals within 24 h. On
the following day, a transparent mass protrudes, which is flat and trian-
gular and has no pigment. This is the *blastema* from which the amputated
part will be rebuilt. After several days (only 3 with certain species) this
blastema shows the first signs of differentiation: as of this moment it is
called the *regenerate* (fig. 3A). The brain, the eyes, the anterior end of the
median cecum and the auricles appear in succession, following a sequen-
tial order that can vary according to the groups. The reconstitution of an
amputated region by a blastema next to the stump is called *epimorphosis*.

If instead of amputating the planarian as described above, one cuts it
behind the pharynx, epimorphosis produces a cephalic region and nothing
more. To reconstitute the intermediary parts that are still missing, the
stump stretches out lengthwise and a new pharynx appears (fig. 3B). This
localized tissue metamorphosis in the vicinity of the cut surface is called
morphallaxis. Similar phenomena are observed when the posterior end is
cut off. The postpharyngeal region is first reformed by epimorphosis. As
for the pharynx, it appears in the stump tissue which was restructured by
morphallaxis into a pharyngeal region. However, if the cut is made closer
to the head, the pharynx appears in the blastema itself. Regeneration is
equally possible along longitudinal cuts, essentially by epimorphosis (fig.
3C).

Excepting the small quantity of tissues produced by epimorphosis, the
regenerating planarian has not really compensated its deficiencies. *It has
reorganized itself completely* and has found a harmonious form. But this
took place *without any appreciable growth* so that the original size of the
individual is not regained at the end of regeneration.

Each time one removes repeatedly the regenerates of fragments, the
animals reform them again and again, becoming in the process smaller
and smaller. The experimenter must sometimes give up because the planarian is so tiny that it cannot be operated upon any longer.[1]

1   The author has isolated pieces of *Dugesia subtentaculata* representing about 1/5
of the prepharyngeal zone, that is less than 1/10 of the total length of the worm [un-
publ. data]. Each week, they were cut in half and each time they regenerated their
head within the same time. After the 4th regeneration, when worms, though perfectly organized, were reduced to less than 1/80 of their initial volume, operations
could no longer be carried out by normal means.

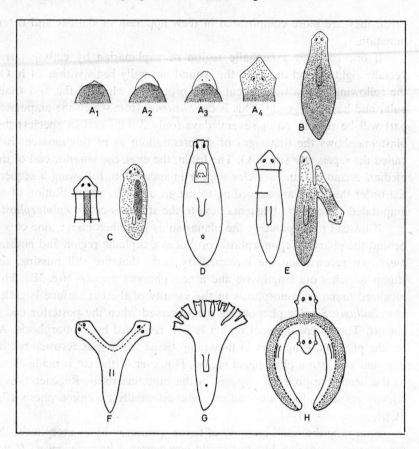

Fig. 3. Various types of regeneration. *A* Head regeneration in *D. subtentaculata* amputated behind the eyes. 1, 2 = Blastemas of 2 and 3 days, respectively; 3, 4 = regenerates of 5 and 9 days, respectively. *B* Morphallaxis in a postpharyngeal fragment of *D. subtentaculata* regenerating the head and the pharynx 2 weeks after amputation. *C* Longitudinal and transversal regeneration in a prepharyngeal piece of *D. gonocephala.* From ABELOOS [1930]. *D* Formation of a second head on the posterior edge of a window in *Bdellocephala punctata.* From BRØNDSTED [1969]. *E* Formation of a second head and tail after a transversal notch in *D. gonocephala.* From ABELOOS [1930]. *F* A two-headed *Dugesia* obtained by an anterior split. From HYMAN [1951]. *G* A ten-headed *Dendrocoelum* obtained by repeated splits. From LUS [1924]. *H* Duplicitas cruciata in *Dugesia dorotocephala* obtained by sagittal split. From SILBER and HAMBURGER [1939].

Evidently if a small fragment of planarian can reform a normal organism, this is due less to the proliferation of undifferentiated material than the extreme plasticity of the tissues and organs. This property which makes both epimorphosis and morphallaxis possible is encountered also in other experimental conditions. The first of these is *intercalary regeneration* (fig. 41). The two ends of a planarian are joined together and the missing parts inbetween will be reconstructed. Grafting of extracephalic and caudal regions into a complete planarian sets off a similar phenomenon, for example a head implanted in the postpharyngeal region induces the appearance of a second pharynx (fig. 39). In certain cases, intercalary regeneration takes place only at the expense of undifferentiated material, as in epimorphosis; in other cases, it implies the reorganization of one of the adjacent parts, i. e. morphallaxis.

Reduction of size as a result of prolonged starvation also illustrates the plasticity of tissues. If we disregard the disappearance of the genital apparatus [VANDEL, 1921a], the worm's morphology remains normal, even after several months of starvation. They can be reduced to $1/100$th of their initial size [PRENANT, 1922] and the numbers of cells decreases accordingly [ABELOOS and LECAMP, 1929; COHEN, 1939].

Morphallaxis is accompanied by a veritable physiological rejuvenation of the tissues. Thus, the planarians are, according to DALYELL's famous expression 'immortal under the edge of the knife'. This capacity is exploited under natural conditions in the form of asexual reproduction of some species and the continuity of species that have lost the capacity of producing germ cells (e. g. *Dugesia subtentaculata*). Individuals that reproduce by fission usually divide behind the pharynx. Morphallaxis which takes place in each of the two pieces probably postpones senescence [CHILD, 1913, 1914]. In an American species, *Planaria velata,* studied by CASTLE [1927, 1928], the detached zooid becomes encysted and undergoes extreme dedifferentiation – probably closely related to starvation. When the zooid emerges from the cyst, the only remaining trace of organization is a difference between dorsal and ventral epidermis. Nevertheless, a normal worm is reformed. One must point out, however, that if tissue modifications are too frequent and especially if no food is given in the meantime, the reorganization is no longer possible. Carried to the extreme, there is no regeneration and finally death results. Thus, planarians that undergo too long a fast can no longer regenerate, and become like those which have been submitted to X-ray treatment [WOLSKY, 1935]. They will die when they are fed again. Likewise, repeated regeneration in large sized fragments cannot go on indefinitely at short intervals.[2]

The constancy of the regenerative processes appearing after an amputation must not lead to a finalistic concept claiming some 'need' or 'desire' to return to the normal organization. Regeneration is a blind mechanism automatically set off whenever, after a wound, the tissues do not succeed in joining up with those from which they have been separated. Then can epimorphosis reproduce a part of the structures of the stump and therefore morphallaxis aggravates the splitting in two. The experimenter can produce to his fancy all sorts of monsters, which no other group of coelomates can provide (fig. 3). Thus, after a transversal notch, the posterior edge of the cut develops a head, the anterior edge a tail. Finally all the animal's structures are doubled. Likewise, the anterior and posterior edges of a hole punched out in the prepharyngeal region develop a tail and a head, respectively. When a sagittal cut is made starting from the posterior end, a two-tailed worm results through transversal regeneration. If the split is extended to the cephalic region, an additional head can also appear. A sagittal cut starting from the anterior end results in two heads. By repeating this operation on each new head, one finally obtains a fan-shaped series of heads on a single tail (fig. 3G). The capacity of planarians to regenerate in all conditions can sometimes result in the worst aberrations under relatively simple operational conditions that the experimenter is usually unable to control (fig. 4). For example, when a fragment is punched out in the middle of the worm, it may happen that many heads develop on both sides of the opening [HULL, 1938]. If the fragment is reimplanted *in situ* with an inversion of its anteroposterior polarity and the graft does not heal everywhere, quite a number of cephalic and caudal ends as well as pharynges will appear [SUGINO, 1938]. With a sagittal incision that goes up as far as the prepharyngeal region, one can obtain a head with inverted polarity against the stump's head. If one splits the two caudal ends once again, a whole new series of supplementary abnormal heads appears [SILBER and HAMBURGER, 1939].

In planarians the doubling of the individual is frequently produced spontaneously as a result of a failure in the determination of the blastema

2   These limits were recently investigated by NENTWIG and SCHAUBLE [1974]. Animals decapitated as soon as the auricles have redifferentiated regenerate 40 successive times if they are fed and only 16 times if they are not fed. However, the time needed to regenerate is remarkably constant, 7 days if the animals are not fed, 6 for the others, as well as in the case of controls regenerating for the first time.

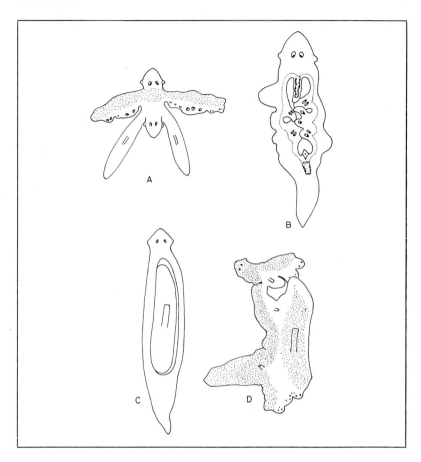

Fig. 4. Anarchic head and tail regeneration. *A* After longitudinal splits in the tails of a *duplicitas cruciata*. From SILBER and HAMBURGER [1939]. *B* After cutting out a central window. From HULL [1938]. *C* When a fragment is punched out and replaced with inverted polarity and does not unite completely with the rest of the body. From SUGINO [1938]. *D* Original.

(fig. 5). In certain species and in certain experimental conditions, a regenerate formed on a transversal cut is the opposite of what is expected: another a head instead of a tail, or vice versa. This is called *heteromorphosis* (LOEB coined this term in 1891 to designate regeneration that does not conform to the organ removed; later, in 1904, MORGAN restricted the term and used it only for cases of inverted polarity). Tail heteromorphosis is only found in marine species. Heteromorphosis usually takes place only

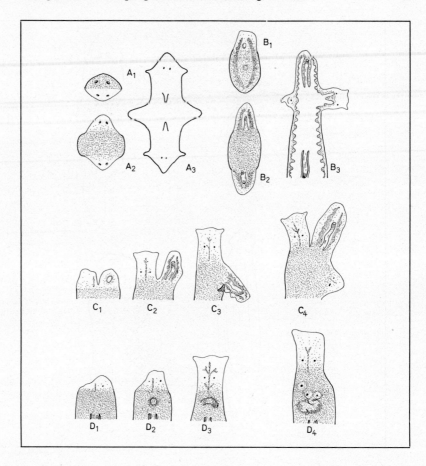

Fig. 5. Abnormal regeneration. *A* Heteromorphic heads. 1 = Pseudoheteromorphosis; 2 = polar heteromorphosis; 3 = *duplicitas cruciata* formed by regulation. *B* Heteromorphic tails. 1 = Pseudoheteromorphosis; 2 = polar heteromorphosis; 3 = *duplicitas cruciata*. *C* Hemiheteromorphosis. 1–3 = Three steps of hemiheteromorphic regeneration; 4 = formation of a supplementary head. *D* Supplementary pharynx. 1–3 = Three steps of the abnormal regeneration; 4 = formation of supplementary ganglia and eyes. A1, A2 from ABELOOS [1930]; A3 from CHILD [1941]; B–D from CHANDEBOIS [1957].

on short pieces that produce two identical regenerates, like mirror images of each other. This is called *polar heteromorphosis*. This anomaly can also occur in isolated end pieces which are duplicated by the regenerate that they form. In certain species, this phenomenon definitely takes place at certain levels. ABELOOS [1932] called this *pseudoheteromorphosis*. A heteromorphic regenerate can also appear beside a normal regenerate on the same cut surface. These are *hemiheteromorphoses* [CHANDEBOIS, 1952, 1953]. After the formation of a heteromorphosis, a certain degree of re-organization of the stump can be observed. The regenerate completes it-self by morphallaxis, then the type of end region that has not been re-generated appears in the area where the two halves of opposite polarity meet. Thus, after head heteromorphosis, the production of one or two supplementary tails has been described; after tail heteromorphosis, one or two supplementary heads appear. An additional head can also appear after the production of a hemiheteromorphosis in which the tail has the inverted polarity. Doubling of the organization is more or less apparent after the regeneration of asymmetrical heads in *Procerodes lobata* [CHAN-DEBOIS, 1957a]. Here morphallaxis leads to an additional pharynx in the prepharyngeal region and sometimes supplementary brain and eyes are also produced.

When a species has any exceptional capacity, it is almost certain that it will also have its 'Achilles heel'. Individuals are threatened by accidental death, which is all the more meaningless because they survive in condi-tions that would be fatal for other species. Among coelomates, there are very few forms that can survive being cut into dozens of pieces. Surpris-ingly, however, the adult planarian, which is rejuvenated by just such an operation, often multiplies to form dozens of normal worms by such pro-cedure. Yet the same species hardly survive a tiny wound. It is practically certain that even though healing has taken place within a few hours the worm will fall ill a few days later. Certain parts of its body will disinte-grate; the epidermis opens in many places; it even develops undifferentiat-ed tumors that finally burst but do not heal. Such a specimen loses its capacity to regenerate if cut. The animal generally dies after agonizing for several days [MARTELLI and CHANDEBOIS, 1973; CHANDEBOIS, in press].

This rapid survey of the principal aspects of morphogenesis in adult planarians allows us to pose the essential problems that will be dealt with in this monograph. First of all, we must ask, where do the cells that build up the regenerates come from and what tissue properties permit the re-organization and the rejuvenation of the stump during morphallaxis? First we will discuss histogenesis along these general lines. Once these questions have been answered, one must proceed to the organic level. How do cells that are used by the blastema manage to reconstitute the structures of the missing parts or, accidentally, those of the opposing extremity? How is tissue transformation directed so as to complete the regeneration or increase a morphological anomaly? These problems relating to regenerative morphogenesis will be discussed in the second part. To conclude, we will report briefly the work done in our laboratory, dealing with the disease mentioned above which is brought about by wounds, since this brings additional information that is needed to understand histogenesis and morphogenesis. This work also poses some questions directly related to cancer.

Histogenesis in Regenerating Planarians

The Current Neoblast Concept

The idea that embryonic cells (neoblasts) exist in the parenchyma of planarians and the migration and aggregation of these cells build up the blastema is inconsistent with certain facts and cannot give a satisfactory explanation of planarian regeneration in harmony with current theories of cell differentiation.

The origins of the current neoblast concept can be found in the last century. At that time, the first histological examinations revealed the presence of large free cells, with a narrow ring of basophilic cytoplasm, scattered about in the parenchyma. After amputation, these same cells were observed accumulating near the surface of the wound or moving towards it by amoeboid movements (fig. 6A). Numerous mitoses were the proof of their multiplication throughout the entire body. These simple observations led inevitably to the idea that regenerates are made of mobilizable embryonic cells, that are kept in reserve in the parenchyma. These cells have been given different names: *'Bildungszellen'* [VON WAGNER, 1890], *'Stammzellen'* [KELLER, 1894], *'Ersatzzellen'* [FLEXNER, 1898], *'formative cells'* [CURTIS, 1902]. Their currently accepted name *'neoblasts'* comes from BUCHANAN [1933]. RANDOLPH [1892], used this expression for large dissepimental cells in the freshwater oligochaete *Lumbriculus,* which are still considered to be responsible for the regeneration of mesodermal tissue [ABELOOS, 1965]. The variety of terms used to name the same cell type is in itself sufficient proof of the disagreement of writers as to their origins and role. The fact that the name 'neoblast' has finally been accepted does not mean that the controversy is over.

At the present time, the majority of writers have accepted the neoblast theory formulated most clearly by WOLFF [1962] and based essentially on the work of his school. According to their findings these free cells have a very high RNA content for they are strongly stained by pyronine [PEDERSEN, 1959; LENDER and GABRIEL, 1960]. They are still in an embryonic state and are stored in the parenchyma where they can be found at every stage of development [LE MOIGNE, 1969]. These cells *alone* make up the regeneration blastema [C. SENGEL, 1960]. After a clean cut, with or without loss of tissue, the neoblasts are mobilized throughout the parenchyma. Attracted from a distance by a hypothetical necrohormone freed by the injured cells, they migrate towards the wound, covering, if necessary, all the distance that separates the caudal end from the cephalic region [DUBOIS, 1949]. To differentiate, they must, on the one hand, divide at least once [DUBOIS, 1949], and on the other hand aggregate. Neoblasts can also revitalize X-irradiated tissues or organs, especially testes [FEDECKA-BRUNER, 1965].

The problem of the origin of regeneration cells was more recently reexamined by BRØNDSTED [1969]. Far from limiting himself to one single theory, he objectively summarized all the works dealing with histogenesis during planarian regeneration, pointing out their weaknesses or subsequent contradictions. Unlike WOLFF, BRØNDSTED is wary of a definitive conclusion and emphasizes that the problems are still far from being solved. He feels, however, that the idea of regenerative reserve cells must be maintained, otherwise one must acknowledge the intervention of dedifferentiated cells, a fact that, for him, has not been proven for any animal group.

BRØNDSTED was right in placing the question over and above all the specialists' differences of opinions, within a more general conception of the mechanisms of cell differentiation. However, ideas have evolved since the time when dedifferentiation was considered impossible, or only limited to a temporary suppression of observable structures. We know now that, in the adult, undifferentiated reserve cells are not the only possible source of new cells. Dedifferentiation has been formally recognized, even in vertebrates [HAY, 1968a]. Moreover, we know that it can be followed by the appearance of other types of differentiated cell in place of the dedifferentiated ones. The classic example is the lens regeneration in adult newts [YAMADA, 1967]. When the crystalline lens is removed from the eye the iris cells on the pupillar margin begin to dedifferentiate. After a phase of activation, during which the synthesis of nucleic acids increases, they redifferentiate again with the help of the eye cup, producing a Wolffian lens regenerate. More recently, CARLSON [1972] has definitely established that cartilaginous nodules appear in muscle tissue when minced muscles regenerate. They come from the redifferentiation of fibers of the Achilles tendon. The discussions over these cases have, of course, no direct bearing on the properties attributed to planarian 'neoblasts': dedifferentiation and the use of elements still in an undifferentiated state are both theoretically possible.

Yet, the essential problem has never been clearly stated: the problem of mechanisms that would make the 'neoblasts' migrate towards the wound, differentiate and rebuild organs, all at a given signal. This problem has to be faced. The current concept involves disorder; the dissemination of 'neoblasts', their individual migrations and their haphazard regrouping. To this, the rigid patterns in the healthy and regenerating animal offer a sharp contrast. Patterns are maintained despite tissue renewal and in the regenerate they reappear according to a well-defined spatiotemporal sequence. It is hard to imagine that such order can derive from chance processes.

Even though the organization of an individual, at every stage of its development, can be considered as the transposition of the linear sequence of genes to a three-dimensional shape, ontogenesis is not directly controlled by genes alone. The DNA during replication and transcription is no more than a 'robot-slave'. For working correctly, it must be externally controlled, without any mistakes. The extrachromosomal environment 'pushes all the right buttons'. At a given moment, it filters genetic information and fixes the rhythm of replications in any cell, differentiated or not. The extrachromosomal environment of the cell is itself conditioned by the extracellular medium. Information available at a given moment in the cytoplasm depends upon previous metabolic activities set up by inductions or by hormone action. This cytoplasmic information can be considered as the *summation of extracellular information* received by ancestor cells. The activation of genes, according to a particular sequence for each cell lineage – that progressively leads to definite differentiation – follows the same general principles. The switching on of a gene activity is necessarily followed by the appearance of new substances in the cytoplasm. The result is the switching on of other genes and consequently the modification of extracellular information provided to the neighboring cells. This again leads to the activation of new genes, and so forth. The cell that remains undifferentiated is obviously suspended within a nucleoplasmic balance. *Only a change in the extracellular information can restore the imbalance that is indispensable for bringing about definitive differentiation.* In certain cases, changes in the humoral interrelation will intervene. In the embryo primordia of organs were determined and these reach full differentiation only during later stages of development. Obviously, the diffusion of a hormone or some other substance (such as a 'necrohormone') cannot be responsible in tissue repair. This is not compatible with the idea of a reserve supply of regenerative cells. The mechanism of keeping a reserve supply of cells for tissues which undergo regular renewal and replacement is known for chick and mouse epidermis [MCLOUGHLIN, 1963]. These tissues degenerate if cultivated alone, after removal of all traces of connective tissue, because the cells of the generative layer become keratinized. Obviously, under normal conditions, the full differentiation (keratinization) of cells of the germinative layer is inhibited by the underlying connective tissue. After cell division, one of the daughter cells escapes and differentiates, even without the help of extracellular information, for its morphogenetic fate has been determined as epidermis. The daughter cell that is attached to the basement membrane remains undifferentiated (nonkeratinized). Therefore, it is unlikely that *generative cells can be kept in reserve unless they receive a definite place during ontogenesis and remain in this place.* In fact, the existence of sparse germinative cells dispersed at random has not yet been definitely demonstrated in any tissue undergoing periodic renewal. The germinative cells usually form the exterior layer (e. g., in bone, thymus, lymph nodes, adrenals, testis) or they are grouped at the apex of a tubular structure (e. g., in the hepatopancreas of crustaceans). They can also form groups that can be detected by their mitotic activity [BURNETT, 1967]. In contrast to this, migrating cells in the adult are incapable of changing metabolic activities, either because their differentiation is irreversible (blood cells) or because they are cancerous – but in this latter case they bring about anarchy in the organization. No valid arguments can be found among these examples to justify the 'neoblast' concept. Only tentative explanations can be offered about how the 'neoblasts' are kept in reserve at the beginning of development, how they migrate without reacting to the various cells they pass by, how they wait until they have aggregated in a sufficient quantity to start differentiating. In place of this uncertain and complicated explanation, other alternative solutions can now be offered. When a differentiated cell suddenly loses the environment necessary for its specific metabolism, it momentarily stops its metabolic activity, turns to increased synthesis of nucleic acid and then redifferentiates in accordance with its new situation.

15 years ago we applied to regenerating planarians PRENANT's method of smears [1922] which he used in his studies on the parenchyma of Platyhelminthes [CHANDEBOIS, 1960]. It became immediately clear that one cannot attribute histogenetic potencies to the 'neoblasts' and that the problem of histogenesis must be reexamined and reinterpreted. After a great

many observations using both light and electron microscopy and experiments involving freshwater planarians, the author believes that a new concept of the histogenesis in planarians during regeneration can be developed. According to this concept, the formation of the blastema, as well as the process of morphallaxis, is due to a *dedifferentiation* process affecting several cell types, especially muscles. This dedifferentiation is accompanied by the temporary appearance of an interstitial syncytium which some writers have already observed. The so-called 'neoblasts' are cells with a short life span. Their morphology and their properties are the same as those of lymphocytes, as PRENANT already postulated in 1922. When these cells undergo cytolysis, they release nucleic acids which they have produced in large quantities and which serves now as a 'raw material' for regeneration processes. This new concept is not only a synthesis of personal results, but also a synthesis of many observations and experiments described in papers by others, both believers and critics of the 'neoblast' concept. In particular LINDH [1958] has already anticipated the trophic role of 'neoblasts', possibly related to immunological functions.

After careful observations of the parenchyma's histological modifications and study of the variations of mitotic activity at the beginning of regeneration, LINDH came to the conclusion that the 'neoblast' concept is not all-encompassing. He offered the hypothesis that these cells are differentiated cells with a very short life span. There is no better introduction to the first part of this book than the following lines, taken from the conclusion of LINDH's article: 'If neoblasts were of greater importance than a *transport and defence mechanisms*[3]; it would be difficult to answer why *all neoblasts do not change in character* and participate in blastema formation. In fact it is only the initiation of regeneration where some neoblasts possibly can be thought to furnish the wound border with repairing tissues ... Instead of *speculation* about neoblasts as repairing tissues I think it will be fruitful to regard the neoblasts as a carrier mechanism, *differentiated for rapid cell division, including a highly effective synthetizing apparatus.'*

The Interstitial Syncytium

The cellular material of the regenerate originates in an interstitial syncytium, produced by the dedifferentiation of certain cells near the cut. The emergence of a

3   Italics mine.

blastema is not due to the accumulation of still more undifferentiated syncytium, but to the beginning of differentiation of the first syncytial units produced in the stump parenchyma.

Light Microscopy

Although most workers in the field have observed the accumulation of free cells near the cuts (fig. 6A) and were convinced that these cells took part in histogenesis, their descriptions of the young regenerates are not all concordant. According to some of them [especially BUCHANAN, 1933; DUBOIS, 1949; see BRØNDSTED, 1969, for an extensive review] there are only regenerative cells in the blastema. C. SENGEL [1960] maintained that the blastema only includes 'neoblasts' enveloped in a very thin epidermis. Other writers, on the contrary, have described the young blastema as a syncytial mass from which the epidermis and the other structures are rebuilt [LANG, 1912; BARTSCH, 1923; STEINMANN, 1926; KIDO, 1961a, b]. These observations are reinforced by STEINBÖCK [1963, 1967] on Acoeles where undifferentiated material is manifestly syncytial.

Such important disagreements in observations can be explained. The constitution of young regenerates of freshwater planarians is very difficult to appreciate on histological sections. As a result of fixation, the newly formed parts are compressed between the reformed epidermis and the stump's parenchyma where the free cells accumulate. Their nuclei are faintly stained by hematoxylin [as KIDO emphasized in 1961], and their cytoplasm is highly basophilic. Some writers who observed the syncytial nature of the blastema [LANG, 1912; BARTSCH, 1923] attributed its formation to the fusion of regenerative cells. However, KIDO [1961a, b] admits this transformation only for the 'neoblasts' that he believes are originating from the nerve cords. Part of the syncytium seemed to KIDO formed directly by the dedifferentiation of the atrium and of gastrodermal cells. However, CASTLE's observations [1927, 1928] already showed that the syncytium is made of totipotent elements. The American planarian *P. velata,* as already mentioned, has a special mechanism of asexual breeding. The zooid has practically lost its organization when it hatches from the cyst. Inside the epidermis there is only an undifferentiated syncytium and no free cells. This syncytium reconstitutes the organs of the adult worm.

The existence of a syncytium has been observed not only in regenerating planarians. When PRENANT [1922] dissociated the parenchyma of non-regenerating specimens with enzyme digestion, he isolated barely vacuolized cytoplasmic masses in which there were several clear round nuclei. Because they looked undifferentiated, PRENANT considered these syncytial

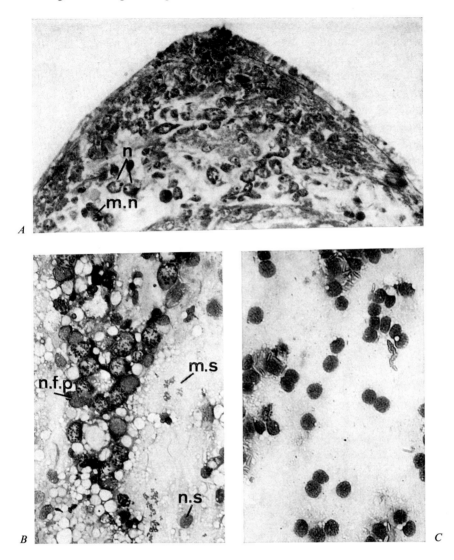

Fig. 6. The parenchyma at the 4th day of head regeneration (first regeneration). *A* Histological section in *D. lugubris.* Susa fix., Masson's Trichromic. ×290. *B* Smear of the stump's parenchyma near the wound after colcemid treatment *(D. subtentaculata).* Mitotic cells are type I cells ('neoblasts'). ×450. *C* Undifferentiated syncytium: smeared blastema of *D. subtentaculata.* ×450. m. n = Mitosis in a 'neoblast'; m. s. = mitosis in the syncytium; n = 'neoblast'; n. f. p = nucleus of the fixed parenchyma; n. s = nucleus of the syncytium.

masses to be young fixed parenchyma that produces connective fibers when differentiating. SEILERN-ASPANG [1960b] mentioned that tumors formed in the epidermal layer of planarians were syncytial.

Even though undifferentiated syncytium has been described on numerous occasions, only two writers have observed that its cytoplasm takes the place of the intercellular ground substance between cells, which is totally lacking in planarians. MURRAY [1927] described this fact very succinctly and made a drawing reproduced here (fig. 7). In spreading cultivated parenchyma, the matrix of individualized cells is obviously of a protoplasmic

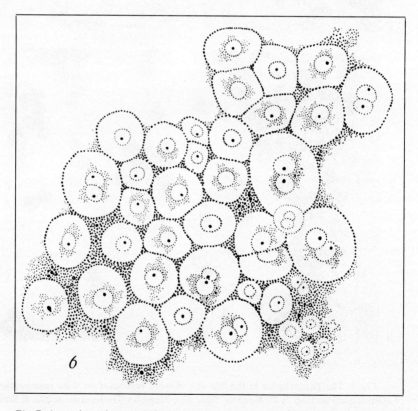

Fig. 7. A portion of a parenchyma outgrowth from an explant in Locke solution modified 20 h after explantation. The granules of the matrix have Brownian movement, and sometimes nuclei may be seen lying free in this matrix, a fact which would suggest that it is of a protoplasmic nature. Figure and legend from MURRAY [1927].

nature because it includes some nuclei and granules with Brownian movements. When HAUSER [1971] studied the first stages of regeneration in the land species *Geoplana abundans,* he noticed the appearance of a syncytium formed by the accumulation of a material coming from the extracellular matrices of the parenchyma. Conscious of the scepticism with which these observations would be received, HAUSER insisted that they could not be attributed to a faulty technique.

There are several reasons why research on the role of a syncytium in the formation of a blastema was discredited: the frequency of artefacts in histological preparations, the difficulty of correctly fixing the planarians' parenchyma, the interest given to regenerative cells rather than to the tissues that enclose them. And yet descriptions of the formation of a syncytium seem to be in fact much closer to the truth than was previously assumed.

The most detailed study of the parenchyma with light microscope was done by PRENANT [1922]. It was essentially carried out by a smear technique. The present author used this same technique to study the parenchyma of regenerating planarians of the species *D. subtentaculata* [CHANDEBOIS, 1960, 1962]. Controls were later carried out on other species: *D. gonocephala, D. lugubris, D. tigrina, Polycelis cornuta, P. nigra, Dendrocoelum lacteum* (unpublished).

A sample is taken for observation and is dried on filter paper. Numerous openings are made in the epidermis, then the sample is slowly pulled, without pressure being exerted, along a slide with a razor blade. The preparation is immediately fixed by drying it with a hair dryer, then it is stained using PAPPENHEIM's panoptic technique (May-Grünwald-Giemsa) for blood smears.

By this technique, the blastema and the stump tissues can be fixed and separately examined. They are completely different in appearance (fig. 6B, C). The young blastemas' parenchyma (almost liquid and easily spread out on the slide) looks like a clear, continuous cytoplasm, that has, in places, absolutely no inclusions. It contains unusually numerous big nuclei, all alike, which do not take up stain easily. They are round, with an inconspicuous membrane and evenly dispersed chromatin. Usually there is a faint nucleolus. In the older regenerates the nuclei have still the same aspect, but the cytoplasm is more basophilic and vacuolar.

The stump tissues in the vicinity of the cut are characterized by a great number of free cells having a very high nucleoplasmic ratio and an extremely basophilic cytoplasm. Their mitotic activity is considerable. Obviously these cells are the 'neoblasts' of various authors. Nothing indicates that they join together to form syncytium of the regenerates. On the contrary, nuclei identical to those of young blastemas are found in numbers in smears taken from tissue neighboring the cut. Surrounding cytoplasm seems to be nonexistent (fig. 8A). Since the smear is spread out on the slide, its basophilic character shows up poorly. Evidently these are the undifferentiated elements to which PRENANT attributed the formation of connective fibers. There are numerous mitoses among them in tissues neighboring transections (fig. 8B–D). Their chromosomes, especially during metaphase and anaphase, are big, faint and more or less sticky. Authors have never observed mitoses in the syncytium on histological sections of regenerating planarians. Therefore, many assume that the chromosomes observed in smears are artefacts caused by the smear technique: 'neoblasts' would tend to burst when dragged along the slide [BRØNDSTED, 1969; PEDERSEN, 1972]. This objection cannot be taken into account because no cellular debris is observed in the vicinity of these mitoses surrounded by an easily spread out syncytium. Moreover, the chromosomes do not look like those of free cells which are short and easily stained. It is clear that the mitoses in the syncytium are quite different from those of free cells, since they are not stimulated by the same factors (p. 62). Histological sections of fragments cultivated *in vitro* show less contracted tissues with mitoses between the free cells. BARTSCH [1923] probably described just such structures, naming them 'extracellular chromatine'.

To prove the relationship between the syncytium of the stump and of the regenerate, one first had to be certain that these particular mitoses really are syncytial. With this in mind cytophotometric analyses were made of the DNA content in the nuclei of the syncytium [CHANDEBOIS, 1973c].

D. *subtentaculata* was analyzed on the 4th day of regeneration. The parenchyma is spread out, then fixed, hydrolyzed and stained with Schiff's reagent, using DECOSSE and AIELLO's method [1966] for blood smears. The interphasic nuclei of the syncytium are easily distinguishable on the preparations. Most of them are found in parts of the smear distinct from the stump tissues. Many among them are very light and they are of the same size, much bigger than nuclei of differentiated cells. Some are slightly bigger and denser. As for the mitoses, Schiff's reaction, more so than Pappenheim's, underlines the differences between mitoses of the syncytium and those of the free cells.

In the samples studied, about half of the interphase nuclei of the syncytium have the same DNA content. The others, especially the bigger and the darker ones, have higher content which, however, never surpasses

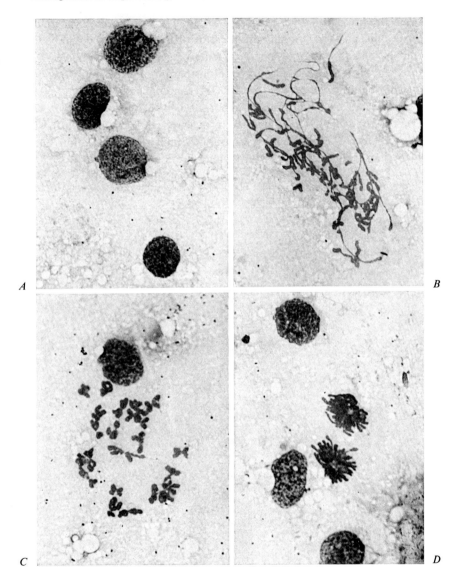

Fig. 8. Syncytium at the base of a 4-day-old regenerate of *D. subtentaculata*. *A* Interphasic nuclei. *B* Metaphase. *C* Anaphase. *D* Telophase. In B and C notice that chromosomes are pale and sticky. Smears. May-Grünwald-Giemsa. ×1,500.

twice the previous value. Most of the mitotic nuclei in the syncytium are far too dispersed to be analyzed, therefore only two counts could be taken. But the results were the same as for the bigger interphasic nuclei. There-fore, the syncytium provides a typical histogram of a tissue in which nuclei are in great mitotic activity (fig. 9). One can conclude that about 50 % of the nuclei are in the G1 phase, about 20 % in G2 phase in which all the DNA is replicated. The others are in the S phase, during which the DNA content progressively increases. Thus, there is good indication that the syncytium really proliferates during regeneration.

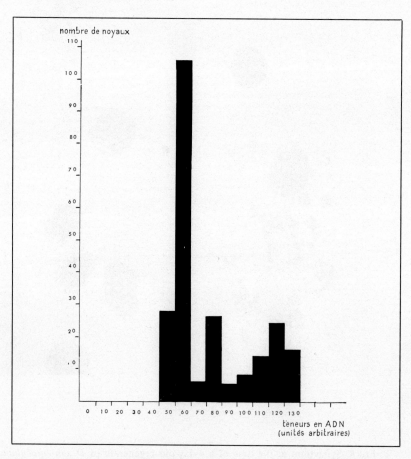

Fig. 9. Variations of DNA content of the syncytium's interphasic nuclei in *D. subtentaculata.* From CHANDEBOIS [1973c].

The hypothesis of the transformation of 'neoblasts' into syncytium – as in fact the whole 'neoblast' concept – can be applied only to worms regenerating for the first time. Unpublished results of the author obtained on specimens that had been made to regenerate repeatedly indicate that the concept cannot be used in repetitive regeneration. *D. lugubris* specimens were used for these studies, since this species has 'neoblasts' that stand out particularly well on histological sections. In the pieces taken from the anterior third of the prepharyngeal zone, the regenerates were removed each week. Each time they reconstituted themselves within the same lapse of time. Samples were fixed from the 1st, 2nd, 3rd, 4th and 5th regeneration, always taken on the 4th day after amputation so as to be sure that they can still form a blastema within the normal time lapse. The numbers of 'neoblasts' is much decreased at the 2nd regeneration and at the 5th one, there are practically no 'neoblasts' left (fig. 10).

If the syncytium is really undifferentiated material, one should be able to follow its redifferentiation in the regenerate. Unfortunately it is usually difficult to identify cell types by light microscopy, and consequently impossible to follow their morphological transformations. Despite the great number of papers dealing with this subject, one actually knows very little about the successive steps of differentiation in any cell type, except in the most recognizable tissue, the epidermis. DUBOIS [1949] and later SUGINO *et al.* [1969] thought that the epidermis is reformed by 'neoblasts' slipping over the blastema surface, although authors such as BARTSCH [1923] and STEINMANN [1926] previously claimed that the new epidermal layer is syncytial. The same observation has been made by KIDO [1961a, b] and more recently by HAUSER [1971] the latter observing the successive steps of healing in the land planarian *G. abundans*. According to these authors, 24 h after amputation an integument already forms over the wound. Its syncytial constitution is still observable at 72 h when the basal membrane is completed and the epidermis becomes progressively cellular.

These seemingly contradictory observations are both partially true and can be reconciled with each other. The author observed the progression of healing with electron microscopy on *D. subtentaculata* [unpubl. data]. The old epidermis reestablishes epithelial continuity; cells near the cut edge elongate and migrate by their pseudopodia on the naked parenchyma. This process of healing, usual for metazoans, was also demonstrated by HIRN [unpubl. data] for the marine planarian *Cercyra hastata* (fig. 44) and by KRITCHINSKAYA and MALIKOVA [1969] for *D. tigrina*. Epithelial cells show no signs of mitoses; all writers agree on this. Therefore, the re-

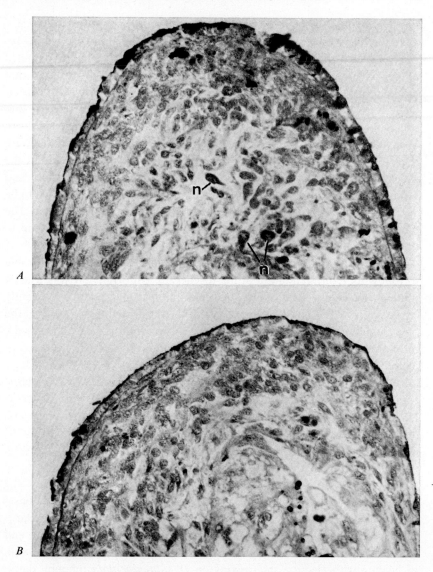

Fig. 10. Disappearance of neoblasts after successive regenerations *(D. lugubris).*
A Second regeneration; neoblasts are few and at a distance from the regenerate. *B*
Fith regeneration; there are no 'neoblasts' although the regeneration was completed
at the same rate as the first one (compare with fig. 6A). Susa fix., Masson's Trichromic. ×280.

establishment of a prismatic epithelium implies that it is enriched by cells coming from the parenchyma by a process most likely analogous to renewals in a normal worm. According to SKAER [1965] the already differentiated cells in the parenchyma cross the basal membrane and embed themselves in the epithelium. This is probably the same phenomenon that DUBOIS [1949] observed and interpreted as the building up of the epidermis by 'neoblasts'. On the other hand, in some explants of D. gonocephala cultured in vitro [CHANDEBOIS, 1968a] syncytium formed tumor-like excrescences with numerous mitoses and without free cells. Their surface was covered with a poorly differentiated epidermal layer; there was no basal membrane between the epidermis and the underlying syncytium, but there was a ciliated border as found in a normal epidermis. In this integument, where nuclei are large and well spaced, there are no cell borders. Therefore, a certain incomplete differentiation does seem to occur when in contact with the medium but in spite of this the syncytial nature of organization is maintained. It is thus clearly indicated that in D. gonocephala the epidermis can be formed entirely by the syncytium. This observation confirms HAUSER's [1971]. It is not unlikely that in certain species the formation of healed epidermis is entirely assured by dedifferentiated material of the parenchyma, while in others by migration of partly dedifferentiated epidermal cells.

Other results which corroborate the origin of differentiated cells in syncytium have been obtained with germinal cells. PRENANT [1922] believed that these cells originate from free cells which exhibit a 'germinal' and undifferentiated chondriome. Today it is generally believed that they differentiate from the 'neoblasts'. Indeed, when planarians have been partly irradiated and recovery occurs, 'neoblasts' are accumulated all around the ovaries [DUBOIS, 1949] and testes [FEDECKA-BRÜNER, 1965]. However, on smears of sexually active worms [unpubl. data, obtained on D. lugubris, D. gonocephala and D. lacteum] the formation of germinal cells is evident and it is easy to follow the characteristic events of meiosis (fig. 11). Oocytes at the zygotene stage are indeed free cells with sparse basophilic cytoplasm. But they do not originate from 'neoblasts', because the leptotene stage is found only in nuclei belonging to elements which are not apparently distinct from the syncytium close to the oocytes. The same observation has been made on the male germ line.

Many authors report that organs or parts of organs disappear during regeneration and that their cells are used as raw material for new histogenesis. This contradicts the neoblast concept which in its rigid form pos-

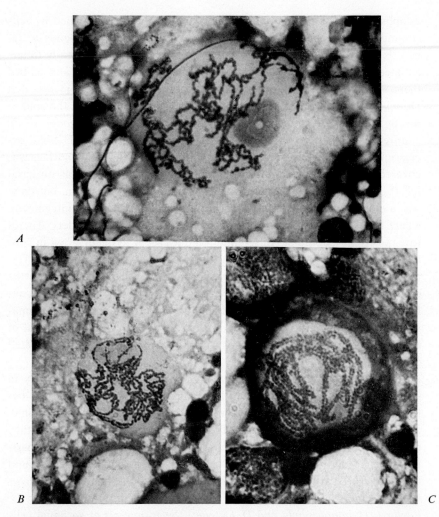

Fig. 11. Three steps of oocyte's growth in *D. lacteum. A* Leptotene stage: the cell is not distinct from the surrounding parenchyma. *B* End of leptotene stage. *C* Zygotene stage: the basophilic cytoplasm is sharply distinct from the neighboring parenchyma. ×3,200.

tulates that cells involved in blastema formation originate exclusively from a stock of undifferentiated reserve cells and this stock could never be enriched with additional dedifferentiated cells [BRØNDSTED, 1969]. These undifferentiated cells are supposed to come directly from embryonic blastomeres [PRENANT, 1922; LE MOIGNE, 1969].

The role of dedifferentiation in regeneration is well documented by the study of variations in the activity of acid phosphatases. Although PEDERSEN [1972] was unable to detect such variations with cytochemical techniques, AUTUORI *et al.* [1965] have shown that the activity of these lytic enzymes increases after amputation and reaches its maximum 2 days later. The same results were also obtained by AUTUORI *et al.* [1965] for *Hydroides norvegica*. They claim that in this annelid, the acid phosphatases take part in the dedifferentiation, which is easily observable morphologically by light and electron microscopy. But for planarians, the same authors attribute the variations to the use of food. Later GABRIEL [1970] also stated that changes in phosphatase activity has nothing to do with dedifferentiation although no further evidence was offered.

In spite of the difficulty to obtain histological proof, many authors think that differentiated cells dedifferentiate into free cells and redifferentiate into another cell type during regeneration. This assumption is based on the fact that while some organs disappear, the number of free cells markedly increases. These results seem to contradict the existence of a syncytium produced by dedifferentiation. Among organs that disappear in the stump at the beginning of regeneration, the copulatory apparatus has been most closely studied, especially by VANDEL [1921b]. Since no phagocytes are involved, it is assumed that this is a case of true dedifferentiation. As the number of regenerating cells at the same time increases, it is logical to assume that these cells are the product of dedifferentiation of the copulatory organs and participate in the reconstitution of other organs, for example the pharynx. GAZSÓ [1958] confirmed this conclusion, adding connective tissue cells to the list of cells that can dedifferentiate. If one believes that free cells are regenerative cells, it is logical to conclude that dedifferentiation leads to their production. But other observations suggest that dedifferentiation and the production of free cells are *two distinct phenomena* stimulated by the same factor: amputation. Using various techniques other than mere amputation, one sees 'neoblasts' disappear in great numbers or completely and, at the same time, substantial dedifferentiation to occur. In such cases the specimens that are able to recover later their normal organization, parenchyma is progressively converted into an undifferentiated syncytium which resembles the parenchyma in the young blastema. This result can be obtained quickly without fail by successive regenerations. The fixed parenchyma in the stump of these very small but still regenerating planarians is completely syncytial and there are very few or no 'neoblasts' in it. Starvation is another method for inducing dedifferentiation without stimulating the supposed regenerative cells [SCHULTZ, 1904]. When planarians are starved for 2 or 3 months, the number of free cells – including 'neoblasts' – far from increasing as one would expect, is markedly decreased, as shown by counts performed by COHEN [1939]. This may explain why VANDEL [1921b] claimed that dedifferentiation is not as clear during starvation as during regeneration.

Syncytium that finally remains can redifferentiate. In *P. velata,* after a complete dedifferentiation has occurred in zooids, there remains only a syncytium with few nuclei, from which new structures are reformed [CASTLE, 1927, 1928]. In smears, the parenchyma of planarians starved for 2 or 3 months strikingly resembles the syncytium of young regenerates [CHANDEBOIS, 1962]. In spite of the complete disappearance of free cells, these worms are again able to recover a normal organization when they are fed.

More recently, WOODRUFF and BURNETT [1965] also described the transformation of gastrodermal cells into 'neoblasts' and interpreted this as a process of dedifferentiation. Their study has been completed by ROSE and SHOSTAK [1968]. Gastrodermal cells were labelled with stained inclusions coming from dyes previously added to the food. Labelled cells were found later throughout the 4-day-old regenerate. Since these authors, referring to LENDER, suppose that the blastema is only formed from 'neoblasts', they conclude that the latter includes dedifferentiated gastrodermal cells. However, these experiments cannot be considered as a clear-cut demonstration of dedifferentiation into 'neoblasts' – or into any type of nondifferentiated material – since there is no proof that the dye is really included in the cells of the regenerate and has not left the gastrodermal cells after its ingestion. However, gastrodermal dedifferentiation has been reported by other authors [KIDO, 1961a, b; KRITCHINSKAYA and MALIKOVA, 1969]. In the next section, when an interpretation of the exact role of the so-called 'neoblasts' in regeneration will be offered, one will better understand that their increased production after traumatization, without being directly related to histogenesis, blurs out the picture of the fixed parenchyma. Because of their accumulation, it is more difficult to observe dedifferentiation that results in the formation of a syncytium and the appearance of mitoses in this syncytium.

In summary, the material used for histogenesis most probably represents dedifferentiated cells near the cut surface. These cells form an interstitial syncytium whose nuclei divide by mitosis. It must be emphasized that the very localized origin of this material was demonstrated by

FLICKINGER [1964], who implanted a piece of tissue, marked with $^{14}CO_2$, into normal planarians. These worms were then decapitated in front of the graft but no radioactivity was found in the ensuing blastema.

Electron Microscopy

PEDERSEN [1961] was the first to study parenchyma by means of electron microscopy in nonregenerating planarians and to prove that it is not syncytial. He observed that its fixed elements are cellular and form very thin and complicated processes. Intercellular spaces are never wider than 200 Å so that there is a nearly complete absence of ground substance except in the basal membrane of the epidermis. This observation explained why collagen was not found in the parenchyma by PRENANT [1922].

We continued the study of the parenchyma of intact planarians and were able to confirm PEDERSEN's observations: the organization appears entirely cellular and there is practically no intercellular ground substance. However, keeping in mind the possibility of dedifferentiation, one can believe that a reserve of undifferentiated elements is not necessary for tissue renewal and regeneration. The replacement of destroyed cells could take place with the dedifferentiation of certain cells forming a transient syncytium, followed by immediate redifferentiation. In a parenchyma containing mainly differentiated cells that are closely packed and whose contours are extremely complicated the rare nuclei of this transient syncytium may easily escape observation by electron microscopy.

As to the occurrence of the process of dedifferentiation, not all writers agree. LE MOIGNE et al. [1965] studied the ultrastructures of blastemas and of tissues in the immediate vicinity of the stump. They described cells that they considered to be differentiating 'neoblasts' and claimed that they found no sign whatsoever of dedifferentiation in other tissues. PEDERSEN [1972] is in complete agreement with these authors. However, he observed disorganized muscle fibers. Because he did not observe sufficient quantities of lysosomes he interpreted this as a degeneration leading to the death of the cells. Only HAY [1968b] is positive that dedifferentiation is a part of regeneration in planarians. She noticed at the base of the blastema the signs found during limb regeneration in amphibians, especially swelling of the nuclei with large nucleoli and loss of cytoplasmic specializations.

During the author's first observations with electron microscopy on planarians regenerating for the first time, it was realized that the ultrastructure was too complicated and could not be interpreted without risk of error [CHANDEBOIS, 1973a]. This has led to the preliminary observations

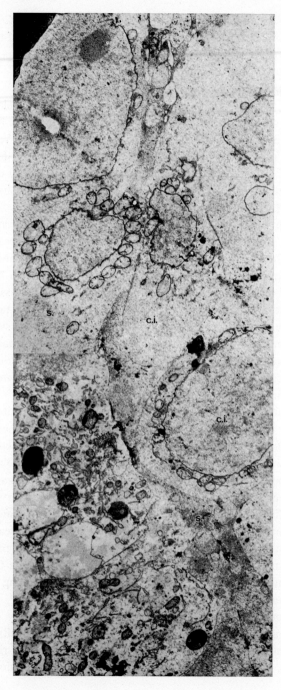

Fig. 12

of samples where dedifferentiation was pushed to an extreme (by prolonged starvation or by successive regenerations). It was also evident that currently used fixations do not show up the very complicated membrane
patterns. KMnO₄ was used first, despite the risk of artefacts. Afterwards
normally fed planarians regenerating for the first time were studied to see
whether the images previously observed were related to the particular physiological conditions of the individuals. To eliminate the artefacts, they
were fixed with osmium or, better, by triple fixation: paraformaldehyde–
glutaraldehyde–osmium.

In sections of regenerating fragments, it is possible to localize precisely the level of amputation (fig. 12). In fact there is a clear demarcation
line, quasi rectilinear, between the stump and the regenerate. On the side
of the former, membranes are very abundant, the mitochondria and lysosomes are numerous. On the side of the regenerate, there are large areas
of ground cytoplasm, entirely lacking reticulum and mitochondria, except
in the immediate vicinity of the nuclei. In longitudinal sections, from the
stump to the tip of the regenerate, there are six types of cellular organization that define five zones, more or less clearly individualized (fig. 13).

The most distinctive region is the base of the regenerate. It can be
recognized at first sight by an exceptional density of nuclei that strikingly
resemble those of embryonic tissues. In planarians fixed with KMnO₄,
especially in those regenerating for the second or the third time, membrane patterns appear to be very simple, and two different zones can be
distinguished. The first one (zone 3) is close to the transection level. Nuclei are scattered in the hyaloplasm, which has practically no reticulum.
They are surrounded by a few mitochondria and do not belong to individualized cell territories. The hyaloplasm, in certain cases, is completely
separated from the stump cytoplasm by the plasma membrane of organized cells. Intercellular spaces directly merge into this cytoplasm. Elsewhere
(fig. 12) there is no such boundary line. The hyaloplasm of the base of the
regenerate is in direct continuation of that of the stump. The second zone
in the regenerate zone 4 is more distal. Its distinctive trait is the indi-

---

Fig. 12. Fragment of a planarian *(D. subtentaculata)* regenerating for the second
time: level of amputation. The part on the lower left belongs to the stump, along
with membrane debris and lysosomes. The rest belongs to the regenerate. Its hyaloplasm is a direct continuation of the same in the stump. Note the formation of thin
membranes that in the syncytium (s) become individualized undifferentiated cell territories (c. i). Permanganate × 15,000. From CHANDEBOIS [1973a].

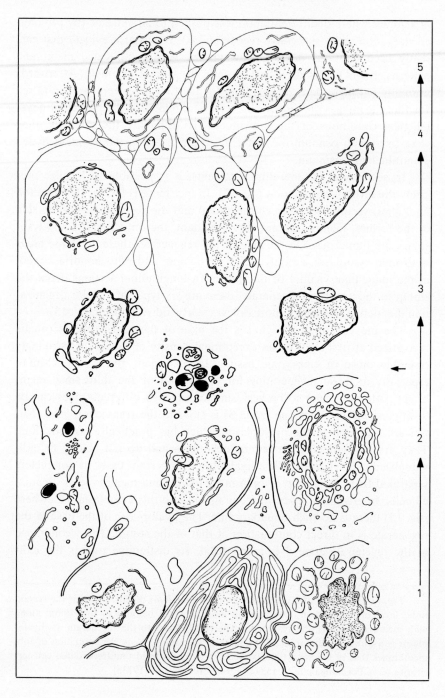

vidualization of cell territories by simple plasma membranes, evidently much thinner than normal cell membranes. They are found around most of the nuclei, at various distances from their mitochondrial aura. They form regular curved lines and delimit extracellular compartments in which the hyaloplasm, having no membranes nor mitochondria, is often electron dense. The new plasma membranes seem to have no connection with the remains of the reticulum which are restricted to the area around the nucleus.

In the stump, close to the transection level, is a zone (zone 2) quite different from both the regenerate (zone 3), where membranes are few, and the normal stump tissue (zone 1), where complicated cells are well separated from each other by spaces of 200 Å, as in nonregenerating worms. At first glance, this zone 2 is distinguished from zone 1 by the enlargement of intercellular spaces. This has previously been observed by LE MOIGNE et al. [1965] and by PEDERSEN [1972]. In planarians regenerating for the second time or later, one observes invariably in the enlarged intercellular spaces scattered round and clear nuclei which resemble those of the blastema (fig. 14). They are surrounded by a few mitochondria and a reticulum which is reduced to tiny vesicles and many ribosomes. These nuclei are separated from the cytoplasm of the neighboring cells by only one single membrane, not by two membranes as in normal tissues. Their cytoplasm is not distinguishable from the material of the neighboring intercellular spaces. In many cases, as in figure 14, the plasma membrane is present only on a part of the cell surface. One also finds cells in which the endoplasmic reticulum is reduced and the cytoplasm is full of ribosomes. In these cells many breaks can be noticed in the plasma membranes. These breaks have been observed after any kind of fixation, and only in dedifferentiated parts. Thus it seems unlikely that they represent artefacts. The enlarged intercellular spaces have not everywhere the same contents.

Fig. 13. Organization of a regenerating planarian and the mechanism of dedifferentiation and redifferentiation. 1 = Stump; differentiated tissues; intercellular spaces narrow. 2 = Tissues next to amputation level; dedifferentiation takes place: regression of endoplasmic reticulum, breaks in and partial disappearance of the plasma membrane; appearance of lysosomes; enlargement of intercellular spaces. 3 = Syncytium constituted of completely dedifferentiated elements, situated at the base of the regenerate; the hyaloplasm is continuous with the material of the stump's intercellular spaces. 4 = Partitioning of the syncytium into undifferentiated cells. 5 = Rebuilding of the reticulum; morphological differentiation is beginning. The horizontal arrow indicates the level of amputation. From CHANDELOIS [1973a].

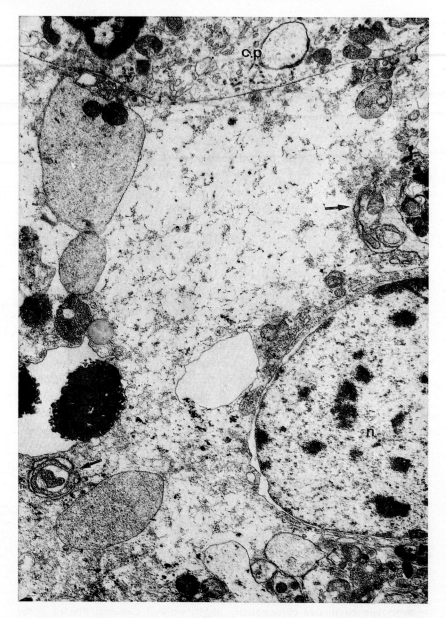

Fig. 14. Release of a nucleus (n) into the intercellular space during dedifferentiation in a planarian regenerating for the fourth time. A large part of the plasma membrane has already disappeared. Remnants of the reticulum, a few mitochondria and many ribosomes are found near the nucleus. Rough reticulum (arrows) and mi-

In some places they look like a homogeneous liquid which locally includes an irregular and loose precipitate, or a dense population of ribosomes (fig. 15, 16). Elsewhere cytoplasmic components are observed: mitochondria, vesicles, rough reticulum. They look either healthy or degenerating. *The most striking feature in dedifferentiating tissues is that the same structures (degenerating reticulum and mitochondria, empty areas or cytoplasm filled with ribosomes) are observed inside and outside the cells.*

Within the dedifferentiating zone, one finds many typical activated nuclei (fig. 17), each with a large nucleolus. The pores of the nuclear envelope are filled with electron-dense – probably nucleolar – material. In the neighboring cytoplasm the old structures (mitochondria, dictyosomes, reticulum) are not completely destroyed. The hyaloplasm in which these cellular debris are found and the adjacent extracellular material are overcrowded with ribosomes.

The conclusion of these observations seems evident. In the vicinity of the amputation surface cells undergo typical dedifferentiation. The first step is the regression of the reticulum and the depletion of mitochondria. The second step is activation: swelling of the nucleus and overproduction of ribosomes. During dedifferentiation, the plasma membrane breaks and progressively disappears, so that the hyaloplasm flows into intercellular spaces. This peculiar process begins in one part of the cell and does not seem to happen at the same moment in every cell type. In some cells it is the first step of dedifferentiation, in others, it follows the outset of activation. This explains why the same structure is observed both inside and outside the cells.

In longitudinal section of regenerates, a gradient of differentiation, decreasing in a basipetal direction, is easily observed. In zone 5 (next to zone 4), all the numerous accumulated nuclei look like those in the syncytium. However, in the cytoplasm, reticulum is more developed and mitochondria are more numerous. At the tip of the regenerate (zone 6), one finds completely differentiated tissues similar to those in normal nonregenerating planarians.

It is now possible to give a precise schedule of regenerative processes. After amputation, the dedifferentiation of the stump's cells near the cut

---

tochondria are in the intercellular space. A cell process (c. p) is separated from the nucleus by a single plasma membrane only. Paraformaldehyde, glutaraldehyde, osmium fixation. ×14,800.

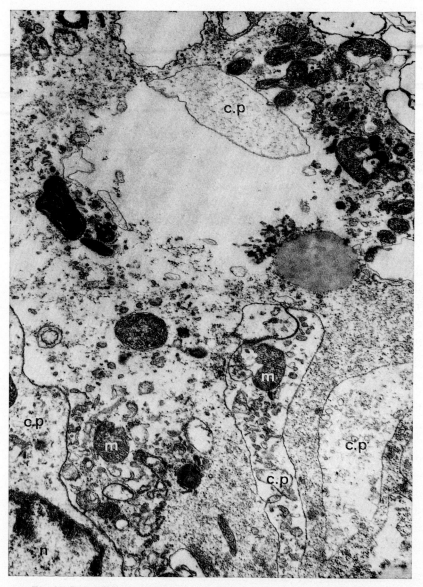

Fig. 15. Intercellular spaces in the dedifferentiating parenchyma of a planarian regenerating for the fourth time. The intercellular material has the same appearance inside and outside the cell processes (c. p). It contains degenerating reticulum, numerous degenerating mitochondria (m) and free ribosomes (on the right). Notice the small gap, a general feature in dedifferentiating tissues, inside and outside the cells. n = Nucleus. Paraformaldehyde, glutaraldehyde, osmium fixation. ×29,300.

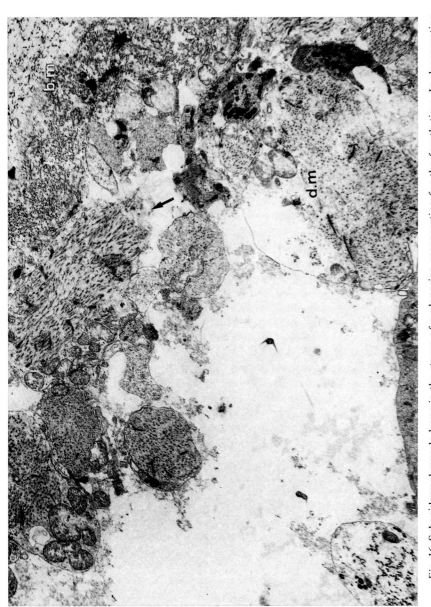

Fig. 16. Subepidermal muscle layer in the stump of a planarian regenerating for the fourth time. In degenerating contractile parts of muscle cells fibrils disappear. In d. m the hyaloplasm has the same appearance as the extracellular material which is abnormally abundant. Plasma membranes (arrow) are lacking in most degenerating fibers. b. m. = basal membrane of epidermis, paraformaldehyde, glutaraldehyde, osmium fixation. ×9,150.

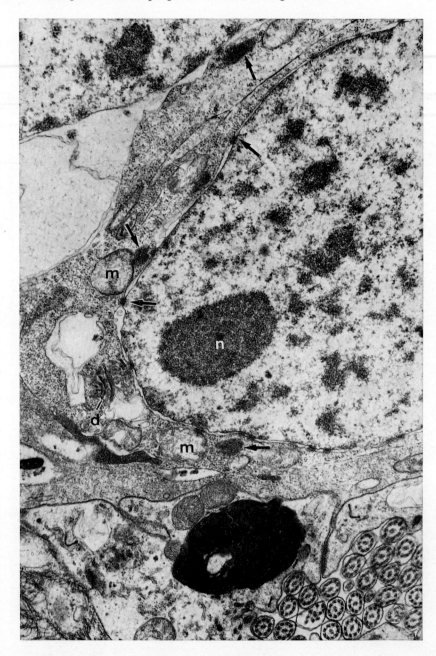

Fig. 17

surface results in the formation of an interstitial syncytium. At the base of the blastema, the syncytium rapidly divides into cell territories that will progressively differentiate into single cells. New undifferentiated cells will in turn become available; they insert themselves between the stump and the cells that are beginning to differentiate. They, too, differentiate, while they are pushed towards the apex of the regenerate by still other, newly formed undifferentiated cells; and the process goes on till the completion of regeneration. Considering the low percentage of units that are found in a pure syncytial state, it is probable that differentiation begins very soon after dedifferentiation. Thus, despite its appearance, at no stage is the blastema formed by an accumulation of completely undifferentiated material.

The author has studied the parenchyma of previously starved planarians to verify whether transformations observed in stump cells of regenerating worms really represent dedifferentiation. In starved planarians, there are almost everywhere cells with broken – or not quite complete – plasma membranes with regressing reticulum and cytoplasm, filled with ribosomes and polyribosomes. There are practically no completely intact cells. Even in the epidermis and the gastrodermal epithelium, the membranes remain intact only in the apical part of the cells and the reticulum is reduced to tiny vesicles. There is a general enlargment of intercellular spaces, in which one finds various cytoplasmic structures in the process of disorganization and sometimes also ribosomes and polyribosomes. Here and there in the parenchyma, there are very large gaps filled with hyaloplasm containing no ribosomes. Their area equals that of several dozen nuclei. The cells that border on these gaps have almost always very incomplete membranes and a more or less reduced reticulum.

The transformations encountered in starved planarians are identical with those found in the stump of regenerating worms near an amputation. However, they are more pronounced, for there are practically no completely individualized cells left and all the tissues have, without exception, very few membranes. On the other hand, there are few lyosomes, so that

---

Fig. 17. Activation following dedifferentiation in the parenchyma of *D. subtentaculata* regenerating for the fourth time. The high activity of the nucleus is shown by the presence of a nucleolus (n) and by an abundant discharge of nuclear material (arrows). This stands in contrast with the neighboring degenerating structures: mitochondria (m), dicytosomes (d). These elements are partly fused with extracellular material lying between them and an excretory cell (below) and with another activated cell (upper left corner). ×22,800.

they do not seem to participate in the destruction of cytoplasmic structures, as they do in regenerating planarians. This destruction seems to be due to the lack of the raw materials necessary for the synthesis of new molecules involved in the renewal processes of the membranes.

In conclusion, one may dare to say that if a planarian could live in a completely undifferentiated state, it would be constituted of a syncytium lacking endoplasmic reticulum. Differentiation is accompanied by the individualization of cells whose plasma membranes are separated by 200 Å spaces. When dedifferentiation occurs in certain cells the plasma membranes disappear. Thus, the nucleus surrounded by several mitochondria and the remnants of the reticulum is found in enlarged intercellular spaces. These spaces do not contain any ground substance but only hyaloplasm and this confirms the observations made by MURRAY [1927] *in vivo*. Together all the dedifferentiated elements constitute an interstitial syncytium. This also confirms the observations of HAUSER [1971] in *G. abundans*.

In the healthy animal, as in the regenerating one, truly syncytial elements are much less frequent than one would suppose from the examination of smears. The young blastema is *not* entirely syncytial. At most only a straight zone at its base is. Similarly, in the stump, interstitial syncytium is found only near the level of amputation. The difference between the results obtained by light microscopy and by electron microscopy, respectively, can be explained by the differences in preparation. In smears the syncytium can only be recognized by the undifferentiated aspect of its nuclei. But nuclei of cells have already this aspect when they are not yet completely dedifferentiated and still retain it when they are already well on the way of redifferentiation.

This new insight in the mechanism of regeneration processes obtained with electron microscopy is radically different from the current 'neoblast' concept. Yet, the latter's strongest arguments come from electron-microscopic studies of the parenchyma. One can attribute this paradox to the fact that an observer is inevitably conditioned by the theory he believes in. For the adherents of the neoblast theory, cells with a 'neoblast' character will be the center of attention. But to those who study electron micrographs from another point of view the picture is more complex, and leads to other conclusions. A comparison of the findings of various authors with electron microscopy is in order here. PEDERSEN [1959] was the first to describe 'neoblasts' in nonregenerating planarians in electron micrographs. He identified them from analogy with their appearance when observed by light microscopy: cells with a high nucleoplasmic ratio, a ribosome-rich cytoplasm, the presence of nucleoli. Their undifferentiated nature is attested by the fact that the endoplasmic reticulum is either absent or very slightly developed.

PEDERSEN's opinion has not been generally accepted because some authors have thrown doubt on the nature of these cells. SKAER [1961] thought that at least some of the 'neoblasts' described by PEDERSEN were glandular cells which had been sectioned near their bases; here cytoplasm is free from inclusions and scanty because it is compressed by the presence of the large nuclei. HAY [1968b] also came to the conclusion that the so-called 'neoblasts' are glandular cells. KISHIDA [1967] thought that the so-called 'neoblasts' are not involved in regeneration at all. He observed that the young eye pigment cells have few structural similarities with them and they rather resemble elements of the fixed parenchyma in the stump. According to him the nature of the parenchyma needs further investigation.

Advocates of the current 'neoblast' concept disregarded these objections, concentrating their attention on the study of the transformations of the 'neoblasts' during regeneration [LE MOIGNE et al., 1965; SAUZIN, 1966, 1967a, b; SAUZIN-MONNOT, 1973; MORITA et al., 1969; PEDERSEN, 1972]. This work has uncovered several interesting points: there are few or no 'neoblasts' in the blastema tissues but they are most numerous in tissue bordering the wound. There they aggregated in small clusters or appear isolated. According to MORITA et al. [1969] their differentiation begins right there. These observations, which we have confirmed, seem to discredit rather than corroborate the 'neoblast' concept. It seems rather unlikely that a regenerate, whose organs must be formed precisely in the right place, at the right time, like organs formed during embryogenesis, would originate from small clusters of cells which were brought together by chance and have started to differentiate *independently from each other*. It is also puzzling that these clusters are never more extensive. On the other hand, it seems that the coordinated process of histogenesis that we have described above explains more convincingly the basipetal progression of the process of regeneration (p. 117).

The details of the description of ultrastructural changes in 'neoblasts' of the stump vary considerably with each writer. According to MORITA et al. [1969], these cells take part in regeneration. Yet, the transformations described as differentiation consist in the accumulation of ribosomes in a cytoplasm where the endoplasmic reticulum is reduced to sparse small vesicles, i. e. symptoms which are considered usually to be the features of dedifferentiation. Disintegration of the Golgi complex is also described, although differentiation would be probably more likely associated with the formation or development of these structures. Finally, the transformations of the 'neoblasts' after amputation strikingly resemble those of any other cell type during early regeneration and starvation. It is, therefore, more likely that the processes observed in the 'neoblasts' during regeneration are in fact symptoms of dedifferentiation. This is corroborated by published electron micrographs showing that the plasma membranes of the 'neoblasts' are often disrupted.

As for 'neoblasts' observed by light microscopy, they belong to a completely different type of cells as will be seen in the text section.

The 'Neoblasts'

The so-called 'neoblasts' are cells with a short life span, specialized in the synthesis of nucleic acids that they release during cytolysis. Their mother cells are probably located in the gastroderm, which explains the influence of feeding on their activity.

Since the dedifferentiation of cells near the site of amputation assures in itself the formation of the blastema, we must ask the question what is the exact role of these free cells which show an increased mitotic activity during regeneration although they are not indispensable to it? Some writers [KIDO, 1961a, b; KRITCHINSKAYA and MALIKOVA, 1969] supposed that 'regeneration cells' act in cooperation with dedifferentiated cells. Writers who have rejected 'neoblast' participation in regeneration believe that these cells bring 'first aid' until the normal mechanisms of histogenesis take over the work. For BARTSCH [1923] these cells are quickly cytolyzed once they have set up the blastema and are replaced by healthy cells. For LINDH [1958] at most they help in healing the wound.

The smear technique finally facilitated the clarification of the role of these 'neoblasts'. On smears of any planarian species, the only free cells which exhibit mitotic activity, and whose numbers are conspicuously increased during regeneration are those which PRENANT [1922] called *type I cells* (fig. 6B). For this reason, the author considers that the type I cells are identical with the cells which are today usually called 'neoblasts' [CHANDEBOIS, 1960, 1962]. PRENANT reached the same conclusion because he observed that their mitochondria belong to what he called the 'undifferentiated type'. When isolated, these cells resemble lymphocytes of vertebrates. They are perfectly round. Their nucleoplasmic ratio is exceedingly high: the nucleus – whose nucleoplasm is strongly basophilic but lacks a nucleolus – is surrounded by a scanty basophilic cytoplasm which forms a regular ring deeply stained by May-Grünwald and hematoxylin.

Basophily of the type I cells is related to their high RNA content as shown by their strong affinity for pyronin that disappears following ribonuclease treatment (the same observation was made in 'neoblasts' by PEDERSEN [1959] and by LENDER and GABRIEL [1960]). Their RNA is synthetized at a faster rate than in any other cell type because they are strongly and selectively labelled after a relatively short period of incubation (24–48 h) with tritiated uridine [URBANI and CECERE, 1964] or tritiated cytidine [CECERE et al., 1964]. The 'neoblasts' have both in resting and in regenerating animals a marked activity of alkaline phosphatase [PEDERSEN, 1959; OSBORNE and MILLER, 1963].

The use of electron microscopy has not yet permitted the definitive identification of these cells. Consequently, we will not enter into the discussion of their ultrastructure here, but restrict ourselves to describing their pecularities as we observed them in smears and histological sections.

Variability in cell size already mentioned by PRENANT [1922] and LINDH [1958], is the most striking pecularity of the 'neoblasts'; it contrasts with the regularity of nuclear sizes in other tissues especially those in the syncytium (fig. 18A). This variability does not depend upon the cytoplasm which is always scanty, but upon the nuclei whose diameter varies between more than the double and less than a quarter of the diameter of the nuclei in the syncytium. This is the consequence of abnormal division processes, easily observed in smeared parenchyma of both regenerating and nonregenerating planarians.

BARTSCH [1923] is the single author who previously described various abnormalities in 'Bildungszellen' accumulated near the site of amputation, inconsistent with their supposed totipotency. He observed cells dividing by unequal amitosis and also nonnucleated fragments. BARTSCH attempted to reconcile these observations with the hypothesis that 'Bildungszellen' are the cells forming regeneration blastema. This is why he supposed that they rapidly divide to act as first aid for tissue repair; after a time they would be cytolyzed and replaced by healthy cells.

Type I cells generally form clusters where they are joined by two, three or more cells. They are not aggregated as might be expected, considering that such an aggregation of the neoblasts would be necessary if they were to be responsible for tissue repair [WOLFF, 1961]. Between two associated cells, there is a very thin and rectilinear septum that is often

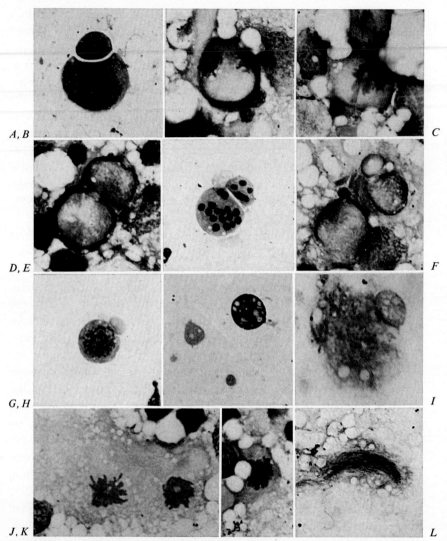

Fig. 18. Type I cells of *D. subtentaculata. A* Two cells of different sizes. *B, C* Asymmetrical telophase. *D* Subequal amitosis. *E* Unequal amitosis in a cell whose mitosis was blocked with colcemid: one daugther cell has 5 chromosomes, the other one 10 chromosomes. *F* Budding: the left cell has a nucleated bud, the right one a swollen cytoplasmic bud. *G* Cell with two cytoplasmic buds. *H* Isolated buds. *I* Budding of a swollen nonnucleated bud. *J* Normal telophase in a residual plasm. *K* Monocentric telophase in a small residual plasm. *L* DNA fibers that are supposed to be remnants of a type I cell nucleus. B, C, D, F, J, K, L = Normal smears; A, E, G, H, I = cells released in a physiological solution. ×1,350.

interrupted medially so that the nuclei are not completely separated (fig. 18D, E). Thus, type I cells seem to divide often by amitosis. The daughter cells are generally unequal and very often to such an extent that one of them may resemble a little bud on the partner (fig. 18F). Type I cells also undergo a 'pinching off' process which liberates minute nonnucleated basophilic fragments (fig. 18G) that can divide and are found in very large numbers in the parenchyma (fig. 18H, I). The high frequency of budding has been corroborated by one observation of the author [unpublished] on hanging drop culture. Many cells attached themselves to the glass cover and were surrounded by a single regular crown of cytoplasmic buds after several hours. It was possible to estimate the rate of this process as approximately one bud per hour.

In type I cells, DNA is replicated at high speed. This is proved by seemingly mitotic activity which is generally very high. But telophases are always abnormal (fig. 18B, C). Cells keep their round shape and the two groups of chromosomes return to the interphasic structure without cytoplasmic division; sometimes the two sets of chromosomes are unequal or gathered in the same half of the cell; they can also be completely fused, the cell still being dicentric (with two centrosomes close to each other) or monocentric (with a single centrosome). Since the mitotic process does not lead to division, the number of chromosomes is doubled and the result is the formation of giant cells. All through the mitotic events, the cells remain marked by their characteristic basophilic cytoplasmic ring, so that at first glance it seems that their nuclear membrane is still present. But when parenchyma is smeared with a drop of physiological solution, some type I cells are liberated. Since they are no longer compressed by neighboring tissues, their cytoplasm spreads on the slide and appears larger and clearer, with no dark staining peripheral ring, and in mitotic cells it is completely mixed with nucleoplasm (fig. 18E). These abnormalities in division processes have in common with endomitosis the production of polyploid cells; but they are not true cases of endomitosis because the nuclear membrane disappears and the mitotic process can reach the telophase stage. Similar abnormal mitotic processes have been observed by Guyot et al. [1960] in corn root meristem after proliferating cells were submitted to high pressures that reduce cytoplasmic viscosity. In type I cells they may also be linked to peculiar physical properties of the cytoplasm. Amitosis can be observed at any stage of mitotic processes.

In the case of D. subtentaculata, not all regenerating worms exhibit the same activity of the type I cells. In some specimens, the parenchyma includes only a few small cells in which very few mitoses are seen. In others, the parenchyma is overcrowded with numerous large cells in which the various mitotic abnormalities and the amitoses can be more easily observed. Similar observations were made in 'neoblasts' of other species by VERHOEF [1946] and by LINDH [1957] who concluded that the number of mitoses is not always high enough to provide cells for a blastema. One can also observe the complete absence of mitoses [DRESDEN, 1940]. This even seems to be the rule for certain species such as the earth planarian of the genus *Rhynchodemus* [BANDIER, 1936]. This variability is probably related to physiological and particularly nutritional conditions. Both aspects of parenchyma can be found in worms of the same batch, but high mitotic rate is more frequent at the end of summer. Type I cells are rare or completely lacking in animals that have been starved for many months [CHANDEBOIS, 1962] or that have performed many successive regenerations (fig. 10). Despite LENDER and GABRIEL's claim [1961] that worms undergoing a 2-month starvation still include many 'neoblasts' most authors agree on the reduction of their numbers or even their complete disappearance after starvation and their reappearance after feeding. COHEN [1939] has followed up this phenomenon most accurately by cell counts. CASTLE [1927, 1928] reported that at the end of their encystment period, the zooids of *P. velata* are reduced to a simple syncytial mass which contains no free cells.

In order to prove that the observed cytological features are not artefacts, regenerating planarians were treated with mitoclastic poisons (demecolcine or colcemide 10^{-4} M) before smearing, so that the chromosomes could be counted in the type I cells. As expected, the normal number of chromosomes was seldom encountered, but a lower number was generally found. Often there were only one, two or three chromosomes in the smallest cells (fig. 22D, F). It is also possible to find large cells containing a number of chromosomes exceeding, or exactly twice the normal. This is proof that certain cells do not divide after telophase but by amitosis. This abnormal mitotic activity of the type I cells can be observed also in normal worms, but it is conspicuously increased during the first day after amputation. It is evident that these cells contribute to regeneration, but not as cells constituting the blastema. In the light of modern cell biology, it is an accepted fact that such strongly aneuploid cells can neither live nor proliferate for a long time, nor form differentiated tissues. LENDER [see discussion in CHANDEBOIS, 1965a] expresses the opinion that aneuploidy is not incompatible with the supposed histogenetic potentialities of the 'neoblasts' because he claims to have found varying numbers of chromosomes even in germ cells.

The fate of the tiny nonnucleated buds can be easily envisaged. They swell while their cytoplasm becomes clear and vacuolized (fig. 18H, I). Afterwards they are always able to send out secondary buds. Finally, the plasma membrane disappears and the cytoplasm of the buds mingles with the surrounding syncytium. Nucleated type I cells of various sizes can exhibit exactly the same swelling and dissolution processes in their cytoplasm. Plasma membranes do not suddenly disappear on the whole cell surface, so that the pinching-off processes and amitoses are still possible after the beginning of cytolysis (fig. 22H). Even mitoses – often with monocentric telophase – can occur (fig. 18J, K). Since the nucleus never becomes pycnotic, chromosome numbers can be counted. They show the same variability as intact type I cells. When the cell outlines can no longer be distinguished, mitoses often resemble at first glance those of the syncytium, but the chromosomes are shrunk and strongly stained (fig. 19B, C, F). They lie in a more or less basophilic and vacuolized cytoplasm. The chromosomes finally become dark thin filaments and eventually seem to disintegrate. Fibrous packs of chromatin resembling locks of hair (fig. 18L) which are found in large numbers in the parenchyma of regenerating planarians might be the last stage of nucleic disintegration. The author has termed the cytolyzing type I cells and the cytolyzing buds 'residual plasms'. During cytolysis, both the cytoplasmic and nuclear components of the type I cells are progressively incorporated into the intercellular material, i. e. the undifferentiated syncytium. This can explain why basophily increases in 'interstitial structures' and why small DNA particles are present in parenchymal cytoplasm, as described by LINDH [1958]. The classical neoblast concept left these phenomena unexplained.

To confirm these observations the DNA content of type I cells was analyzed with a cytophotometer [CHANDEBOIS, 1973c]. On smears stained with Schiff reagent, the interphasic nuclei of these cells could easily be identified by frequent amitoses and by the great variation in size (fig. 19D, E). They are also characterized by their very homogeneous chromatin, which is, after Feulgen stain, much bluer than that of the other nuclei. Altogether 274 measurements were made on cells of various sizes that were not budding or dividing by amitosis. The results obtained are widely scattered (fig. 20). The lowest value of DNA content corresponds to about one chromosome. The largest nuclei, irregularly shaped and often overlapping other cells, could not be fully measured. However, the maximum

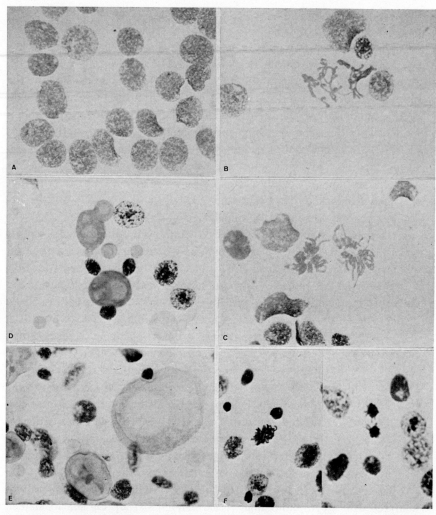

Fig. 19. Comparison between interphasic and mitotic nuclei of the syncytium and those of type I cells after Feulgen stain *(D. subtentaculata). A* Interphasic nuclei of the syncytium. *B, C* Mitoses in the syncytium. *D, E* Various sized nuclei of type I cells. *F* Two mitoses in type I cells. From CHANDEBOIS [1973c].

values obtained equal about six times the DNA content of nuclei in the syncytium. Together with chromosome counts, these analyses confirm the conclusion that the type I cells (neoblasts) are not formative components of the regeneration blastema.

However, STÉPHAN-DUBOIS [1965] and PEDERSEN [1972] express the

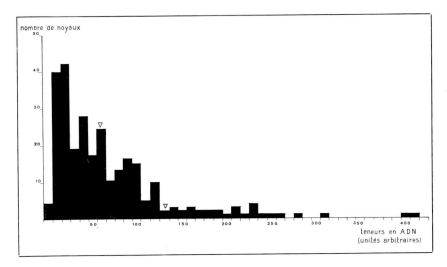

Fig. 20. Variations of the DNA content in interphasic nuclei of type I cells in *D. subtentaculata.* Arrows mark the extreme values found in the syncytium's interphasic nuclei (fig. 9). From CHANDEBOIS [1973c].

view that the type I cells are not identical with 'neoblasts' because with PRENANT's technique (smears), one cannot see the voluminous nucleolus that is so obvious in 'neoblasts' in histological sections. Various authors (p. 56) observed mitoses in 'neoblasts' throughout the parenchyma. On Feulgen stained smears of stump tissues, one finds only one type of mitotic nucleus that has short and very chromophilic chromosomes. They were analyzed with cytophotometry. The measurements were made on 224 nuclei between the metaphase and the telophase. These results were also widely differing from each other and the values were, as a general rule, noticeably lower than those obtained for nuclei in the syncytium (fig. 21). The two extreme values have a 1 to 12 relationship. Considering the results furnished by type I cells during interphase, one would have expected an even greater range. In fact, after Feulgen stain, mitoses with either too many or too few chromosomes, cannot be identified without risk of error and were not analyzed. An examination of the histogram clearly shows that there are no two distinct categories of cells ('neoblasts' and type I cells) as claimed, the former having a steady morphogenetic role in regeneration, the latter being short-lived and similar in their role to some vertebrate blood cells. If this would be so, one would expect at least one peak in the histogram for the cells with a morphogenetic role in regenera-

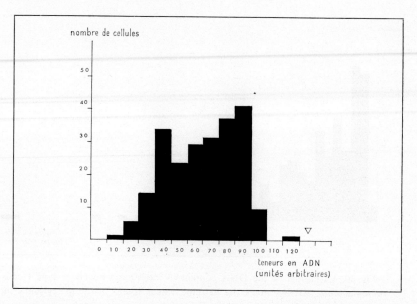

Fig. 21. Variations of the DNA content in mitotic type I cells in *D. subtenta-culata*. The arrow marks the DNA content of the syncytium's mitotic nuclei. From CHANDEBOIS [1973c].

tion, i. e. the supposed 'neoblasts', as they should have a constant full complement of chromosomes. But there is no sign of such a peak in the histogram.

The conclusion that the type I cells are not 'neoblasts' and that there are no 'neoblasts' in planarians is also corroborated by the fact that free basophilic cells are lacking in the parenchyma of the young regenerates up to about the 8th day after amputation. After that time one can observe some round nuclei in the syncytium surrounded by a thin layer of basophilic cytoplasm [CHANDEBOIS, 1962]. This can explain the observation of PEDERSEN [1959], already mentioned, that good preparations of 'neoblasts' for observation with phase contrast microscopy can be obtained only from 6-day-old regenerates and not from very young blastemas. The first mitoses in the regenerate appear with the differentiation of these cells. This observation may explain the observation of LENDER and DUBOIS that 'neoblasts' penetrate one by one into the regenerate and that there are mitoses in the blastema itself [discussion following CHANDEBOIS' paper, 1965]. It is clear that the 'neoblasts' observed in the late regenerates are formed *in situ* like any other cell type when tissues are differentiating.

Since type I cells seem unable to divide by any other process than amitosis and since they are finally destroyed, they are necessarily generated by other cells. Examination of slides during regeneration suggested that the type I cells are produced directly by the syncytium like any other cell type [CHANDEBOIS, 1962]. But a closer scrutiny revealed that throughout regeneration the number of mitoses in the syncytium is 5–10 times lower than in the type I cells [CHANDEBOIS, 1972]. It is, therefore, quite impossible that type I cells, which themselves cannot multiply, would be formed one by one, by the syncytium in which the number of mitoses is insufficient to account for the production of both differentiated tissues of the regenerate and type I cells. Between the totipotent elements of the syncytium and type I cells, there are necessarily specialized mother cells which can still divide. In smears, these mother cells have not been observed. They are probably present in compact tissues which cannot be spread over the slide since most of the type I cells are found near such compact spots. After current histological investigations with *D. subtentaculata,* it seems that type I cells may originate from gastrodermal cells. Other species showed this result much more clearly; detailed observations by WOODRUFF and BURNETT [1965] revealed that the 'neoblasts' can be formed by dedifferentiation of specialized intestinal gland cells, not only during the first 4 days of regeneration, but even in the uninjured worm. LINDH [1958] observed that the larger 'neoblasts' are found close to the gastrodermis. OSBORNE and MILLER [1963] followed alkaline phosphatase changes associated with feeding in *D. tigrina*. One or two days after cooked food ingestion, the activity of this enzyme increased strikingly. It is completely absent by the 8th day but is found by this time in resting 'neoblasts'.

In conclusion it can be said that the so-called 'neoblasts' are not undifferentiated cells (in spite of some morphological characters which would indicate this) but belong to a differentiated system, the *system of type I cells* [CHANDEBOIS, 1970]. This system is replenished by the syncytium during regeneration as any other differentiated tissue. The system (fig. 22) includes mother cells – probably originating from the gastrodermis –, type I cells that synthetize both RNA and DNA at high rates and become smaller and smaller because of repeated amitoses, and residual plasms (swollen type I cells which have lost their plasma membrane and anuclear fragments) which release nucleic acids in the syncytium. The strongest argument for this conclusion is furnished by LINDH [1958] who observed that in regenerating *Euplanaria polychroa,* the largest 'neoblasts' are in

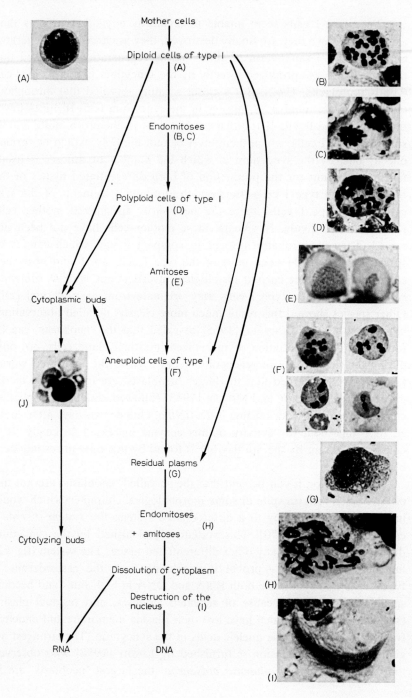

Mother cells

Diploid cells of type I
(A)

Endomitoses
(B, C)

Polyploid cells of type I
(D)

Amitoses
(E)

Cytoplasmic buds

Aneuploid cells of type I
(F)

Residual plasms
(G)

Endomitoses
+ amitoses   (H)

Cytolyzing buds   Dissolution of cytoplasm

Destruction of the
nucleus   (I)

RNA   DNA

(A)   (B)   (C)   (D)   (E)   (F)   (G)   (H)   (I)   (J)

contact with the 'demodulated' gastrodermis where mitotic activity is increased. The 'neoblasts' exhibit a marked size gradient so that the cells and their nuclei at the periphery are the smallest. He also observed that during regeneration, the affinity for basic stains is increased in the intercellular spaces.

The activity of the system of type I cells is deeply influenced by physiological conditions, probably by nutrition, as shown by the sensivity of the system to seasonal rhythm and starvation.

The frequency of 'endomitoses' and amitoses in regenerating and nonregenerating worms, the obvious influence of nutritional factors on their numbers, the cytolysis that finally results, taken all together are good indications leading to the conclusion that type I cells are differentiated in order to play a trophic role in the organism. Since cytological and autoradiographic examination demonstrates a very active synthesis of RNA and especially of DNA that are produced even during cytolysis, it is logical to believe that type I cells release into the syncytium nucleic acids needed for proliferation and for differentiation. Components of plasma membranes, which look very active during the amitoses and the pinching-off process, are probably also synthesized at higher rates than in other cell types. Other substances may be also elaborated and released and may play an important role in histogenesis, but we have no idea about their possible nature. Since the type I cell system is also active in the healthy animal, it must have a role in normal cell renewal too. However, observation of regeneration in fragments where this system had stopped working showed that it is not indispensable for histogenesis; it simply plays an auxiliary role.

LE MOIGNE's [1963] observations with *P. nigra* strongly indicate that the type I cell system is differentiated very early in the embryo. In fact, when vitelline cells start to be digested by the temporary gastroderm, a varying number of cells, showing signs of budding appear in the vitelline syncytium. LE MOIGNE believes that these cells absorb fragments of the syncytium, because if one accepts the idea of a pinching-off process, one also has to accept the fact that the syncytium is an integral part of the

---

Fig. 22. The type I cell system in *D. gonocephala* (2 n = 16). *A* Young interphasic type I cell. *B* Anaphasic diploid cell. *C* Telophase of an endomitosis. *D* Tetraploid cell. *E* Unequal amitosis. *F* Aneuploid cells. *G* Interphasic residual plasm. *H* Mitotic residual plasm with one cytoplasmic bud. *I* DNA fibers that are supposed to be remnants of a type I cell nucleus. *J* Budding cytoplasmic bud.

embryo, although its extraovular origin is incontestable. Two other observations indicate that one can consider these cells as the first budding type I cells. On the one hand, the basophily of the buds is as strong as that of the cytoplasm of cells to which they are attached. It stands out against the syncytium's clear background. On the other hand, LE MOIGNE notes the presence of free buds *after* and not before the stage when budding cells appear. Thus, these cells could already be used to make nucleic acids needed for the growth and the differentiation of the embryo from digested vitelline cells.

Analysis of Mitotic Activity During Regeneration

After an amputation, the syncytium formed near the cut shows mitotic activity varying with respect to the amount of tissue to be reconstituted. In contrast, the temporary rise in the mitotic activity of the type I cells and of residual plasms can be observed throughout the stump and appears to be the reaction of an immunological system to a challenge.

Mitosis in 'neoblasts' can be observed very easily with usual histological techniques; this explains why many experimenters have studied their variations in number during regeneration (almost always regeneration of the head), either by simple global evaluations or by precise counts. All this work has led to the same results. Mitotic activity increases for the first 3 or 4 days and then it decreases [DUBOIS, 1949; McWHINNIE and GLEASON, 1957; WOODRUFF and BURNETT, 1965]. Even though the 'neoblasts' are most numerous near the site of amputation, their mitotic activity is stimulated throughout the parenchyma [STEVENS, 1907; VERHOEF, 1946; BEST *et al.*, 1968; GABRIEL, 1970]. From this it follows that the number of 'neoblasts' produced is more or less proportional to the size of the fragment and, consequently, *inversely proportional to the size of the regenerate,* a situation that is the exact opposite of what one would expect if these cells would be morphogenetic elements. The greatest number of 'neoblasts' is produced in the extreme case of a mere incision without a loss of tissue [DUBOIS, 1949], that is when they serve no useful purpose. Theoretically speaking, one could produce an infinite number of 'neoblasts' by repeated wounding. Except for LINDH [1957] no one has remarked on the contradiction that this extraordinary waste would represent if the 'neoblasts' were regenerative cells, as opposed to the economy of regulative processes both in embryos and in adults.

Most of the studies on mitotic activity after amputation have revealed only a partial aspect of the mitotic activity of the parenchyma as a whole, for they have not dealt with the syncytium or with residual plasms. Moreover, they were carried out almost exclusively on animals amputated behind the head. Therefore, it was necessary to repeat numerical studies on pieces put under different experimental conditions, and on smears rather than sections, in order to get a precise idea of the variations of mitotic activity in residual plasms and in the syncytium.

Counts were taken in smeared tissues of the asexual species *D. subtentaculata* [CHANDEBOIS, 1970, 1972] because in a study of sexual worms, mitoses in germinal cells are hardly distinguishable from other mitoses and would distort the findings. To increase the efficiency of the counts, regenerating fragments were treated 24 h before their fixation with a mitostatic poison (colcemide 10^{-4} M). During 'endomitosis' the type I cells (mCI) are easily identifiable by their spherical form and their plasma membranes that separate them from the neighboring parenchyma. 'Endomitoses' in residual plasms (mPR) at the beginning of cytolysis can be recognized by the basophilic cytoplasm that generally includes many vacuoles and is not clearly delimited from neighboring parenchyma. When cytoplasm is completely dissolved, 'endomitoses' in residual plasms do not resemble mitoses in the syncytium (mS). In the latter there are long, pale and fairly sticky chromosomes; the cytoplasm is hyaline and hardly visible. In residual plasms chromosomes are chromophilic and generally contracted in a vacuolized and basophilic cytoplasm. Since parenchyma is easily smeared, cells are not frequently damaged. The technique has been proved completely reliable by the calculation of correlation coefficients between mCI and mPR and between mCI and mS. They reach unusually high values of 0.80–0.94.

The first experiments were undertaken in order to establish the changes in mitotic activity with time in regenerating fragments corresponding approximatively to the anterior third part of the prepharyngeal region. Two types of fragments were investigated (fig. 23). In one type (exp. 1) the two amputations were made at the same time and regenerates in both amputation sites were building up when fixation was performed. In the other type (exp. 2), an anterior amputation was made first and a posterior one was performed only just before fixation. Thus, a regenerate was formed only at the anterior end. In both series of experiments the average number of mCI increases in the same manner – apparently exponentially – until the 3rd or 4th day and then decreases (fig. 24). In both experiments the variations of average numbers of mPR are about the same as those of mCI. The ratio R/I (numbers of mPR/numbers of mCI) was computed for each day; it remains approximately constant, about 0.30 in experiment 1 and 0.20 in experiment 2 (table I). Thus, the respective mitotic activities of type I cells and of residual plasms vary in the same way during the

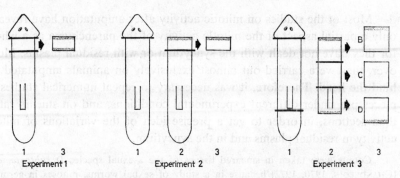

Fig. 23. Preparation of fragments in which mitoses are to be counted. 1, 2 = transection; 3 = constitution of the fragment at fixation; regenerating surfaces stippled. From CHANDEBOIS [1972].

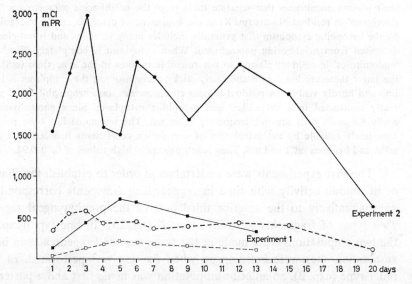

Fig. 24. Variation of mitotic activity in type I cells and residual plasms during the regeneration of fragments of *D. subtentaculata* with one (exp. 2) or two (exp. 1) regenerates. □, ○ = mPR; ■, ● = mCI. From CHANDEBOIS [1972].

whole regeneration. In both types of experiments, the average number of mS first increases in the same manner as that of mCI. This appears clearly when average numbers of mCI and mS from both experiments for the first 4 days of regeneration are plotted in semilogarithmic coordinates. In both experiments the two lines exhibit exactly the same slope. In experiment 2,

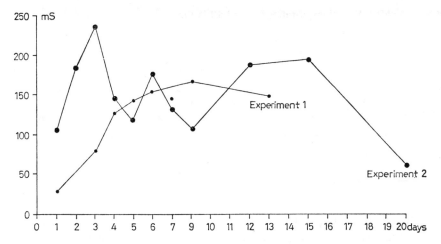

Fig. 25. Variation of mitotic activity in the syncytium during the regeneration of fragments of *D. subtentaculata* with one (exp. 2) or two (exp. 1) regenerates. From CHANDEBOIS [1972].

numbers of mS decrease after the 3rd day in the same manner as in mCI (fig. 25) as shown by the almost constant value of S/I (numbers of mS/ numbers of mCI) which is about 0.07 (table I). But in experiment 1 the number of mS is still increasing after the 4th day. On the 13th day, the number of mS has not decreased significantly (fig. 25). The ratio S/I unmodified up to the 4th day, increases gradually up to the 13th day, from 0.17 to 0.42 (table I). Thus, correlation between mitotic activity and the conditions of the experiment is observed only in the syncytium. The long lasting high mitotic activity in the fragments of experiment 1 is probably related to the large quantity of cells which have to be produced for the posterior wound.

In order to analyze the relations between the three types of mitotic cells, the numbers of mPR and of mS were plotted against the numbers of mCI in logarithmic coordinates. The equations of regression lines for each experiment have been computed. For mPR against mCI (fig. 26), the plotted points show no great variation in either of the two series. The two regression lines are very close to each other. Again we reach the conclusion that the two different experimental conditions had no great influence upon the relation between type I cells and residual plasms. In contrast, the points given by the numbers of mS against numbers of mCI in the two

Table I. Variations of ratios R/I and S/I in time

	Experi-ments	Days											
		1	2	3	4	5	6	7	9	12	13	15	20
mPR	1	0.26		0.33	0.25	0.30	0.31		0.32		0.30		
mCI	2	0.21	0.20	0.19	0.26	0.18	0.20	0.17	0.22	0.19		0.20	0.19
mS	1	0.17		0.16	0.16	0.19	0.22		0.31		0.42		
mPR	2	0.06	0.07	0.07	0.08	0.07	0.06	0.05	0.06	0.07		0.13	0.09

experiments are distributed into two distinct clusters that slightly overlap. The regression line of experiment 1 has about the same slope and is above the one of experiment 2, at a distance of log. 2 (fig. 27). From this analysis, one can deduce that mitotic activity is stimulated throughout the parenchyma in the system of type I cells. It is stimulated only near to the site of amputation in the syncytium so that the number of mS is approximately doubled when the regenerating surface is increased twofold in fragments of the same size.

If this conclusion is correct, fragments taken from amputated worms far from a regenerating surface should show in type I cells and in residual plasms the same degree of mitotic activity as in the region near to the regenerate, whereas the number of mS should remain low. To test if this prediction is correct, another series of counts was undertaken (exp. 3, fig. 23). Worms were transected behind the eyes and at the level of the mouth. These long fragments were allowed to regenerate. Nine days later, just before fixation and after 24 h in colcemide, they were cut again into three pieces, one pharyngeal (D) and two prepharyngeal, the anterior one bearing a regenerating wound (B) like the fragments of experiment 2 and the posterior one (C) without a regenerate. The numbers of mCI and of mPR had the same range in the three fragments of each worm. However, the numbers of mS were much lower in C fragments. The results were plotted in logarithmic coordinates. The relation between mPR and mCI is the same as in experiments 1 and 2. The relation between mS and mCI is the same in fragments B (with one regenerate) as in the ones of experiment 2, but the points obtained from fragments C show a separate cluster (fig. 28). The number of mitoses in the syncytium is five times higher in fragments B than in fragments C. This study confirms that residual plasms

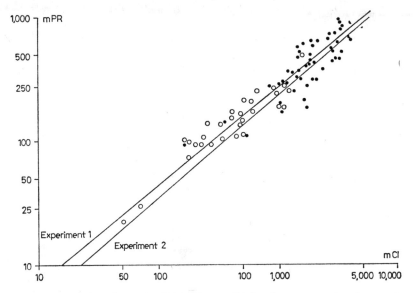

Fig. 26. Regression lines of mPR against mCI for the regenerating fragments of *D. subtentaculata.* ◯ = Experiment 1; ● = experiment 2. From CHANDEBOIS [1972].

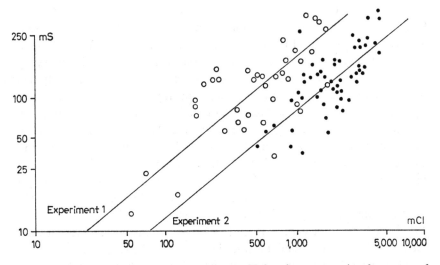

Fig. 27. Regression lines of mS against mCI for the regenerating fragments of *D. subtentaculata.* ◯ = experiment 1; ● = experiment 2. From CHANDEBOIS [1972].

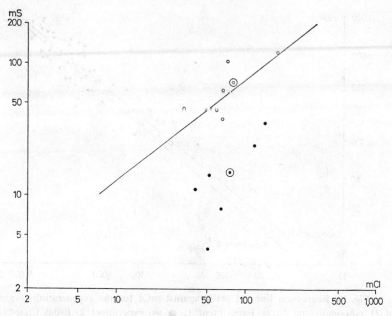

Fig. 28. Comparison between the activities of the syncytium in fragments having one regenerate (○) or no regenerates (●) in *D. subtentaculata.* The regression line corresponding to the mS of experiment 2 has been drawn. Points corresponding to arithmetical means are encircled. From CHANDEBOIS [1972].

and type I cells belong to the same system whose mitotic activity presents neither the same spatial distribution nor the same temporal variations as found in the syncytium during regeneration. Mitotic activity in the syncytium is closely related to regenerative processes. It is increased only near to the site of amputation and lasts longer in fragments that have a greater amount of tissue to regenerate. Mitotic activity in type I cells undergoes variations in the same way as in the so-called 'neoblasts': it increases in the whole worm after any amputation and always in the same manner regardless of the amount of tissue lost. It strikingly resembles an inflammatory process. The factor that enhances this type of mitotic activity is evidently spread evenly over the whole parenchyma. DUBOIS [1949] thought that the factor might be a necrohormone, but LINDH [1957–1958] suggested that water, which is a hypotonic medium, penetrates into the parenchyma and stimulates mitotic activity in the so-called 'neoblasts'.

These results have been corroborated by the fact that analogous activation is observed after injections of various substances (oil, peptones) and even tap water [MARTELLI and CHANDEBOIS, 1973]. Countings were made on histological sections of *D. lugubris* which had been injected with oil. Mitotic activity increases in the whole parenchyma during the first 7 days, then decreases. Other countings were performed on *D. subtentaculata*. Spring water was injected at the postpharyngeal level and 1–6 days after this operation, a prepharyngeal fragment was taken out and smeared. Colcemide was applied 24 h prior to fixation. Other planarians that served as controls were amputated behind the eyes then fixed in the same way. Mitoses were counted in type I cells, residual plasms and syncytium. In the type I cell system mitotic activity increases rapidly in the controls and in injected worms alike and reaches its maximum on the 2nd or 3rd day. In the syncytium, the mitotic activity is stimulated in regenerating controls but remains fairly low in injected animals.

It is a striking coincidence that the system of type I cells reacts like the thymus of the vertebrates. After intraperitoneal injections of peptones in young cats, DUSTIN [1923] observed that the number of mitoses increased rapidly within the next 4 days and reached about 4 times the value for normal thymus. Afterwards it progressively decreased. DUSTIN also observed that during this regression period the system cannot be stimulated again by another injection. He called this phenomenon 'cinéphylaxie'. It seemed worth investigating whether the system of type I cells of planarians exhibits this same property [CHANDEBOIS, 1971b, 1972]. For this purpose, the anteriormost region of the head was removed and the worm allowed to regenerate (exp. 4); 8 days later (4 or 5 days after maximum activity had been reached) short fragments were isolated from the anterior pharyngeal region and smeared 1, 2, 3 and 8 days later, after having been treated with colcemide for 24 h prior to smearing. Arithmetical means have been computed for each day and plotted on graphs in normal (fig. 29) and in semilogarithmical coordinates. The type I cell system is not stimulated a second time. The number of mCI actually decreases after the second transection. However, the syncytium is stimulated so that the ratio S/I, exceedingly low on the first day, increases in later stages. Its mitotic activity increases as in controls during the first 2 days after amputation but does not maintain the same rate after the 3rd day, probably as a consequence of the depressed activity of the system of type I cells.

The stimulation of mitotic activity in the system of type I cells after an injection of diverse substances and the existence of cinephylaxy indicates that this system, as well as playing a trophic role, might play also a part in the defense of the organism, as LINDH [1958] already suspected.

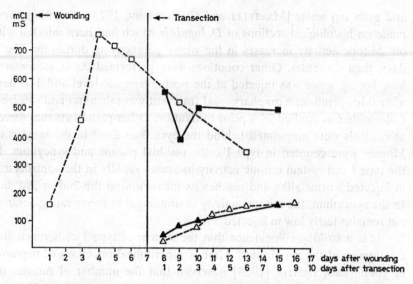

Fig. 29. Variation of mitotic activity in 'type I cells' (■) and in the syncytium (▲) during regeneration of fragments of *D. subtentaculata* wounded 8 days before amputation. Curves of experiment 1 (□, △) serve as controls. From CHANDEBOIS [1972].

To summarize, a wound stimulates mitoses in two distinct cellular systems: the first corresponding to 'neoblasts', reacts as a defense mechanism throughout the stump; the second represents elements that generate blastema and that are formed by dedifferentiation near the site of amputation. COWARD *et al.* [1970] previously postulated this dual response of cells to the same amputation. Instead of counting the mitoses at different levels and at different stages of regeneration, they analyzed the activity of thymidine kinase, related to DNA replication. The latter first increases throughout the stump, but later on the maximum shifts towards the wound. The authors suppose the occurrence of a localized dedifferentiation in the vicinity of the amputation and, at the same time, the mobilization of 'neoblasts' throughout the parenchyma.

Mitotic counts have confirmed that there is no clear relationship between mitotic activity of type I cells and the size of the blastema. This apparently paradoxical phenomenon can be explained by the relation that exists between syncytium and type I cells as suggested by data obtained with electron microscopy. One must remember that the young blastema is not an accumulation of undifferentiated cells. Its appearance coincides with the organization of distal parts of the regenerate by cells beginning to differentiate. Afterwards, the regenerate grows gradually as the cells previously made available for differentiation are incorporated. Therefore, growth of the regenerate depends exclusively upon the process of differentiation. It progresses at the same rate in all regenerates formed at the same level, regardless whether or not the syncytium proliferates by multiplying its nuclei through mitoses. Electron microscopy shows that cells dedifferentiating when there are no type I cells present have both nucleoli and polyribosomes suggesting that they can differentiate even without the help of a trophic system. Since syncytial elements can differentiate without previous division, the participation of mitoses in regeneration appears to be simply a saving in differentiated material. Because there is a mathematical relationship between the mCI and the mS, this means that the proliferation of the syncytium is closely related to the activity of the type I cell system and that the saving in differentiated material is proportional to the activity of this system. One understands that the incidence of the activity of this trophic help on regeneration is not situated at the level of the regenerate itself, but at the level of the stump near transection where dedifferentiation that occurs in tissues is consequently reduced to a minimum, as observed in every regulative process.

X-Ray Irradiation

Planarians amputated immediately after X-ray treatment partially regenerate, even though all mitotic activity is suppressed. The loss of the regenerative power observed when amputations are made several days after X-ray treatment can be explained by the exhaustion of RNA freed by type I cells present at the time of treatment and without which the X-rayed syncytium cannot differentiate.

Since the work of BARDEEN and BAETJER [1904] it is a well-known fact that regeneration of planarians can be inhibited by X-rays. Since the supposed regenerative cells disappear [CURTIS and HICKMANN, 1926], it was logical to believe that they indeed play a role in regeneration. This discovery was the starting point of the studies of WOLFF and DUBOIS [1947] elaborating a technique to demonstrate the migration of 'neoblasts'. Their purpose was to eliminate these cells selectively in one part of the worm and to mobilize healthy ones from the nonirradiated part by amputation. Totally irradiated planarians *(D. lugubris)* were studied as controls for determining experimental conditions in which 'neoblasts' are killed and regeneration becomes impossible. Intact worms that had received more than the threshold dose (between 2,000 and 5,000 r according to the season) produced necroses and finally died. Those amputated *24 h after total body irradiation* formed a little blastema which was quickly resorbed. DUBOIS observed that interphasic 'neoblasts' are destroyed and supposed that the dividing ones are kept healthy and could still form a minute regenerate after transection. The experimental worms were given a lethal dose of X-rays, but the posterior part of the animal was protected by a lead screen. After a while the irradiated zone became necrotic. But when an amputation is performed in the irradiated zone, necroses progressively disappear and regeneration finally occurs. The greater the distance between the site of amputation and the nonirradiated region, the longer is the delay in blastema formation. SUGINO *et al.* [1970] came to similar conclusions by joining halves of X-rayed worms with halves of nonirradiated ones. These data form the strongest support to the 'neoblast' theory.

Since one could suppose that the revitalization of tissues results from a diffusion of substances, this latter technique was perfected by LENDER and GABRIEL [1965]. Pieces of healthy planarians whose 'neoblasts' have been labelled with tritiated uridine were grafted onto irradiated planarians and then the head removed. The head regenerated after about 1 month and the cells of the regenerate were labelled. This experiment permitted the authors to ascertain definitively the existence and migration of 'neoblasts'. It is no wonder that today this concept is generally accepted and dominates the thinking of developmental biologists working with planarians.

Indeed, WOLFF and DUBOIS [1947] have based their work upon a generally accepted hypothesis. They themselves accepted it without question since they only looked for a sound technique for the demonstration of 'neoblasts' migration. Unfortunately the 'neoblast' theory does not completely fit the disregarded observations made by the first authors who tried X-irradiation, although these observations were confirmed by other authors. BARDEEN and BAETJER [1904] have shown that *one must postpone the amputation for about 2 days after irradiation in order to ensure complete blockage of regeneration.* Planarians whose heads are amputated immediately after an irradiation first regenerate more or less normally and later die. WEIGAND [1930] also reported that regeneration with normal differentiation was possible in *P. nigra* when specimens were transected immediately after irradiation. Smaller regenerates were produced in planarians amputated 1 day later, and no regeneration took place in the worms amputated after the 2nd day. VAN CLEAVE [1934] found complete regeneration followed by cytolysis in *Stenostomum tenuicauda* (Archoophores) after irradiation and immediate amputation. It is evident that if the X-ray effects were restricted to killing 'neoblasts', such a regeneration would be impossible – or at least stopped at an early stage – depending on the length of survival of the irradiated cells. More recently, KRATOCHWIL [1962] emphasized the generally underestimated importance of regeneration in irradiated planarians and reached the conclusion that 'neoblasts' are actually not killed by X-rays. He used 'harder' X-rays, so that worms *(D. gonocephala)* amputated immediately after irradiation could not regenerate. However, regeneration was possible when worms were amputated at least 24 h before irradiation and also when they were amputated from 0 to 72 h after an irradiation administered 1–3 days after a previous amputation or incision without tissue loss. KRATOCHWIL's interpretation is that X-rays inhibit a preliminary process, which he called 'activation' and which would be a prerequisite for differentiation; such 'activation' would be stimulated in the whole worm by any incision, with or without tissue loss.

At about the same time *D. subtentaculata* was totally irradiated under the same conditions as in DUBOIS' [1949] experiments and amputated immediately or a few days later [CHANDEBOIS, 1963b]. The results were the same as those obtained by BARDEEN and BAETJER [1904] and WEIGAND [1930]: inhibition of regeneration was observed only after delayed amputation. Sooner or later all the regenerating worms exhibited necroses according to their physiological condition (from 4 days for well-fed worms to 2 weeks for starved ones). The necroses start with a regression of the regenerated head, in the same way as in the irradiated intact controls. In these controls (fig. 30A), the epidermis shrinks first in the auricles which roll up dorsally. It looks like an integument in the process of wound healing and the resulting local contraction makes the pigment condense in dark patches. The same process occurs on the ventral side and then at the tip of the head. The eyecups shrink, then the head is progressively resorbed. The eyespots and the brain are retracted into the prepharyngeal region and the anterior edge of the planarian looks like a healed cut surface, with a black border, that DUBOIS [1949] called 'collier de nécroses'. It probably represents the accumulation of cephalic pigment moved back by the resorbed epidermis. Sometimes the epidermal shrinking starts dorsally in the eyecup region. An identical phenomenon takes place a little

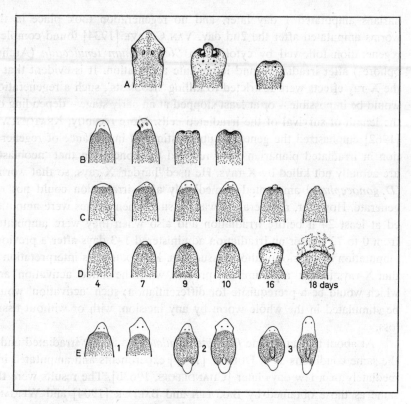

Fig. 30. Regeneration of irradiated planarians *(D. subtentaculata). A* Cephalic regression of irradiated controls. *B* Regeneration of well-fed worms amputated immediately after irradiation. *C* Regeneration of worms starved for 10 weeks and amputated immediately after irradiation. *D* Regeneration inhibited in well-fed worms amputated 48 h after irradiation. *E* Comparison between maximum sizes reached by regenerates in planarians amputated after irradiation. 1 = Worms transected behind the head; 2 = prepharyngeal pieces; 3 = pharyngeal pieces; for each set of drawings, figure on the right, worms wounded 3 days before irradiation; figure on the left, no wounding. A–D from CHANDEBOIS [1963b]; E from CHANDEBOIS [1971a].

later along the edges of the tail, then dorsally along the base of the pharynx. The planarian body is bloated by pressure coming from the internal tissues. The epidermis usually gives in, especially where it is reduced to a membrane, covering a healing wound. In these weak spots, debris of whitish tissues will be discharged. Healing is possible but sooner or later a more extensive rupture takes place and the planarian dies a few hours later. The necroses in the epidermis suggest that its cells are not replaced because cell differentiation processes are blocked.

Histological examination (sections and smears) has shown that the syncytium remains apparently normal and forms new differentiated tissues in the regenerate without an intermediary step of cell multiplication. Mitoses are completely lacking in the system of type I cells as early as the first hours following irradiation. Type I cells vanish within about 2 days. Thus, KRATOCHWIL's [1962] experimental conclusions and DUBOIS' [1949] histological observations, which seemed contradictory, are both correct; the regenerative cells are not killed by X-rays, but the so-called 'neoblasts' are destroyed. On the other hand, KRATOCHWIL's hypothesis is in perfect agreement with the explanation proposed above; inhibition of regeneration can only be correlated with a preliminary process of differentiation which occurs during the first days following an amputation. Since type I cells disappear after an irradiation and are believed to discharge into the syncytium substances necessary for its differentiation, it is logical to assume that their activity is correlated with this preliminary process of regeneration.

Further experiments on *D. subtentaculata,* irradiated with doses of 8,000–12,250 r have confirmed that possibilities of regeneration depend upon the activity of the system of type I cells. When worms are amputated just behind the eyes immediately after irradiation (fig. 30B), heads regenerate almost completely but grow no further after about 10 days. Eyes differentiate 1 or 3 days later than in controls and they lack eyecups because the normal pigment is not replaced; auricles are also not reconstructed. In short fragments morphallaxis is impossible so that regenerated tails always lack pharynges. When worms are transected 24 h after irradiation, regenerates are normally shaped but have small eyes, and remain very short. When amputations are performed 48 h after irradiation (fig. 30D), usually only the wound heals. The wound epidermis is more or less distended because of the swelling of parenchyma, but eyes can be formed. In a few exceptional cases, some worms formed exceedingly short regenerates with one or two eyes. It can be concluded that the longer the amputation is delayed, the smaller will be the amount of differentiated tissues formed. These complicated features can be fully explained by the well-known sensitivity of nucleic acid synthesis towards X-irradiation. In the syncytium RNA production is probably stopped [KELLY, 1961] so that cells which would be available for the formation of the regenerate cannot differentiate. DNA replication is also influenced by X-rays so that mitoses are promptly inhibited and the production of type I cells is immediately stopped. Type I cells finally disappear because they are cytolyzed and are not replaced. But those cells of this type which have been produced just before irradiation are normal and it is possible that the RNA that they release into the syncytium is sufficient for allowing regeneration for a time. When these substances are exhausted, cell differentiation is no longer possible. Regeneration cannot be completed even in the already produced levels (this explains the lack of auricles and of pigmentation); the formation of the more proximal parts is stopped so that morphallaxis is never observed, and epimorphic parts are often shorter than those of nonirradiated worms. Finally, the fact that the regenerates of the irradiated worms are not completed recalls regeneration of starved planarians. This similarity was pointed out by WOLSKY [1935] who thought that irradiation and starvation act upon the regenerative cells in the same way. Indeed, the syncytium is hampered in both cases by the depression or complete elimination of the system of type I cells.

The preliminary experiments described above suggest that, in contrast to nonirradiated planarians, there is a close relation between the length of the regenerate and the number of type I cells which are in the parenchyma right at the time of irradiation, because the RNA produced by these cells may be the only source of nucleic acids necessary for the differentiation of the syncytium. Experimental conditions that stimulate or slow down the activity of the system of type I cells are well-known and, consequently, one can prove this hypothesis [CHANDEBOIS, 1963b, 1971a]. Starved worms include no type I cells, or have very few of them. Such worms survive after irradiation much longer than well-fed worms, a fact that makes one think that undifferentiated material is perhaps more resistant. (Apparently, the resistance of starved worms to X-rays is related to their depressed metabolism and especially their slow rate of cell renewal.) However, when starved planarians are decapitated immediately after irradiation, the heads which they regenerate are smaller than those regenerated by well-fed, irradiated controls (fig. 30C). On the other hand, the number of type I cells can be increased by stimulating the system which produces them; the increase reaches its maximum 3 or 4 days later. Worms can be stimulated to produce more type I cells by cutting the marginal edges of the tail. Such worms were irradiated 3 days later together with uninjured controls of the same batch. In a first experiment injured worms and controls were, immediately after irradiation, either amputated behind the head or dissected into four pieces (cephalic, prepharyngeal, pharyngeal and postpharyngeal). In each type of fragment, the regenerates were larger in the worms previously wounded than in the controls (fig. 30E). The difference was particularly clear-cut in the decapitated worms which produced new heads with auricles and eyecups; in prepharyngeal pieces, the regenerated tails formed in a few cases, a little pharynx; in pharyngeal fragments, heads reached a considerable length. In a second similar series of experiments, amputations and dissections were delayed (fig. 31). Controls amputated 48 h after irradiation had no regenerates or very short ones. Stimulated worms amputated 48 h after irradiation exhibited regenerates of the same size as the planarians amputated immediately after an irradiation; those amputated 3 days after irradiation formed slightly smaller regenerates, though they were still somewhat longer than the regenerates formed by controls amputated 48 h after irradiation. The same results have been obtained with worms whose system of type I cells was stimulated by injection of water [MARTELLI and CHANDEBOIS, 1973]. These experiments clearly show that the size of blastemata and the length of the delay of amputation after irradiation, necessary for a complete inhibition of regeneration, are closely related to the number of type I cells. Cytological observation enables us to give a plausible interpretation to these phenomena. Type I cells formed before irradiation probably continue their normal development up to cytolysis. They release RNA that will be available to the syncytium for a certain length of time, thus permitting regeneration to take place for a certain time, even though transcription is more or less blocked in the syncytium. The variations and results of these experiments recall the work of KRATOCHWIL [1962]. He 'activated' the worms by postcephalic amputation or wounding without tissue loss then irradiated them with a heavy dose that completely suppresses regeneration in unwounded controls. Regeneration was possible in the 'activated' worms if amputation was performed immediately or within 3 days after irradiation provided the worms were 'activated' more than 24 h before irradiation. The best results were obtained when worms were wounded 3 days before irradiation. Recently, ZILLER [1974a, b] also admitted the existence of RNA discharge right after amputation and expressed the view that this could be eventually used by 'regenerative cells' in which RNA synthesis is inhibited by experimental procedures. ZILLER's tool was the antibiotic drug actinomycin D which specifically inhibits RNA transcription on DNA templates. When planarians were placed in an actinomycin D solution for the first 30 h after they were decapitated, their regeneration capacity was noticeably lowered. Around the 7th day, 50 % of the worms had formed

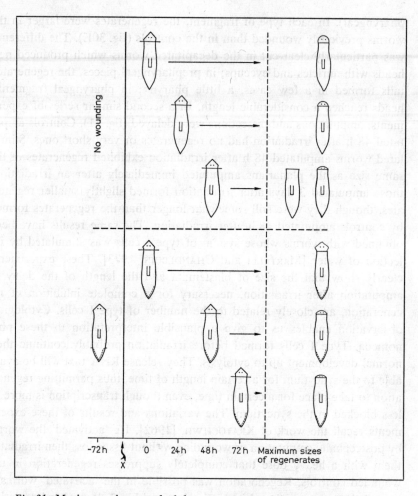

Fig. 31. Maximum sizes reached by regenerates of irradiated planarians according to the lengths of delay of amputation. Regenerates are larger when worms are wounded 72 h before irradiation.

eyes, in contrast to 100 % in the untreated controls. However, if the planarians undergo the same treatment after a second decapitation (removal of a 3-day-old cephalic regeneration blastema), they regenerate like the untreated controls. If planarians are treated with the antibiotic for 30 h immediately after decapitation and are then amputated again 3 days later, they show the same reduced regeneration capacity as animals treated with actinomycin immediately after a single decapitation. The only possible ex-

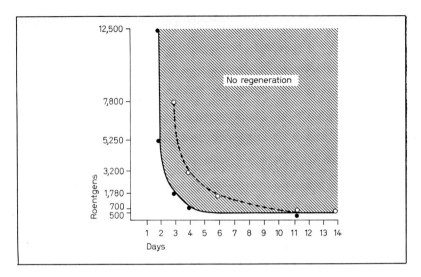

Fig. 32. Variation of amputation delays after irradiation, necessary for a complete inhibition of regeneration plotted against the dose of X-rays (in roentgens). ● = *D. subtentaculata;* ○ = *C. hastata.*

planation for these results is that the first amputation makes RNA available for the syncytium which can be used later. It should be noted that ZILLER, who used the 'neoblast' theory in her early work could not explain the 'activation' phenomena observed in the actinomycin experiments.

The inhibition of mitotic activity by X-rays cannot be the only cause of the inhibition of regeneration, because the delay necessary for a complete inhibition depends on the dosage above the threshold value for mitotic inhibition. In our own work on *D. subtentaculata* (fig. 32), the length of exposure to X-rays (85 kV, 18 mA) was varied and worms were amputated at various times [CHANDEBOIS, 1965b]. Using doses between 700 and 12,250 r, all the worms were finally killed. Mitoses were inhibited with 525 r or more. The lower the dose in roentgens, the longer the time between irradiation and amputation necessary for inhibition of regeneration. With 525 r, regeneration does occur in spite of the lack of mitoses, but is temporarily inhibited when amputations are made 11 days after irradiation and starts only 10 days later. With 350 r, although mitotic activity is not inhibited, regenerates become smaller and smaller as amputations are performed later and later; but regeneration is never completely inhibited, since it could be observed in worms amputated 20 days after irradiation. Regeneration is normal again in worms amputated 1 month following irradiation. The minimal postponement of amputation required for a complete inhibition also varies according to the species. In *P. cornuta,* there is no possibility of regeneration above 9,000 r [CORNEC, unpubl. data]. In *D. lugubris,* regeneration occurs when transections are performed between the 1st and the 7th day after a 10,250 r irradiation [D. GRÉGOIRE, in CHANDEBOIS, 1968b]. In the marine triclad *C. hastata* [HIRN, unpubl. data] the curve has the same shape as the one based on data obtained with *D. subtentaculata* (fig. 32). The minimal postponement of amputation needed to produce inhibition of regeneration also varies according to the nature of X-rays. KRATOCHWIL [1962] irradiated *D. gonocephala* using hard rays (170 kV, 19 mA) with the same quantity of roentgen's (8,000 r) which, at a lower voltage (95 kV) does not inhibit regeneration for 48 h. In KRATOCHWIL's experiment regeneration was inhibited even in worms amputated immediately after irradiation. In short, when doses are progressively increased, there is a decrease in amputation delays necessary for a complete inhibition of regeneration. This suggests that a second effect of irradiation reinforces the result of the primary stoppage of the production of type I cells. This factor may be the destruction of RNA. It is known that RNA is degraded by X-ray irradiation, with different speed, according to the doses, as observed in various systems such as spleen, thymus, bone marrow which synthetize RNA in large amounts [GOUTIER, 1961]. On the other hand, it has to be pointed out that the regenerates formed after amputation performed just after irradiation are still growing while regeneration is already inhibited after delayed amputation. This observation suggests that certain substances are probably protected from degradation once regeneration has started. Thus it is also possible that inhibition of regeneration could be linked to a progressive decrease of an enzymatic activity necessary for the incorporation of substances liberated by the cytolysis of type I cells. Degradation of substances and inhibition of their uptake in the regenerate are not mutually exclusive and might act together in the reduction and disappearance of regeneration potency. As time passes, the available amounts of those substances which are incorporated at the beginning of differentiation processes would be lower, until they are too low to produce even a minute blastema. The delay of the effect of irradiation evidently depends upon the total quantity of substances available immediately after irradiation.

At this point a tentative general explanation of the effects of X-ray irradiation on planarians can be made. X-rays inhibit neither the dedifferentiation process nor the terminal stages of cell differentiation. As in other metazoans, they stop the synthesis of nucleic acid and degrade RNA synthesized before irradiation. As a result, the syncytium normally produced is not activated and is unable to undergo differentiation. It could do so if it were to receive RNA from type I cells. But the inhibition of mitoses curtails the production of the latter. Thus, the syncytium is left only with a limited reserve: the RNA released by type I cells before irradiation. This stock is itself degraded and all the more rapidly as the dosage of X-rays is increased. When this stock of RNA is used up, the possibility of regenerating is completely lost.

It is now possible to propose a new interpretation of the classical experiments which suggested the migration of 'neoblasts' to DUBOIS [1949]. According to this author, an amputation gives a mobilization order to healthy 'neoblasts' of the nonirradiated part of the worm so that they migrate towards the wound. They build one blastema when they reach the wound after having revitalized necrotic tissues on the way. We think that the system of type I cells is only preserved in the healthy zone. After an amputation water freely penetrates that parenchyma and reaches the system of type I cells throughout the worm, but the system cannot react in the irradiated zone. The syncytium formed near the wound cannot immediately receive nucleic acids so that it cannot differentiate. In the same way cell renewal is impossible and necroses appear in the epidermis. However, the system of type I cells can react in the healthy part where there is an overproduction of nucleic acids, as in a normal regeneration. These cells assume the role of progressively revitalizing necrotic tissues and finally lead to the formation of the blastema. Regeneration of irradiated planarians which have received a graft of healthy tissue can be explained in the same manner. The experiments of LENDER and GABRIEL [1965] and GABRIEL [1970], mentioned above (in which they grafted tissues previously incubated with tritiated uridine and recovered labelled material in the regenerate and in free cells between the graft and the regenerate) was considered a hard proof for the 'neoblast' theory. But it can be assumed that if the 'neoblasts' – type I cells in our terminology – finally undergo cytolysis, their labelled RNA will be released and can be incorporated both in cells forming the blastema and in irradiated tissues in which cell renewal is restored. Indeed, the autoradiography of LENDER and GABRIEL shows labelling both in the 'neoblasts' and elsewhere, especially at the level of the graft; radioactive spots are scattered all over the blastema.

In vitro *Culture*

Small fragments put in a nutritive medium which stimulates mitotic activity in the type I cells generally dissociate with no visible proliferation. If the gastrodermal cells swell and obstruct the wounds, the fragments do not disaggregate but they rapidly shrink. The fact that one cannot obtain histotypical cultures from planarian tissue can be attributed partly to the migration of type I cells towards the medium where they cytolyze and partly to the semiliquid consistency of the interstitial syncytium formed by dedifferentiating cells.

Many authors have pointed out that planarian tissue culture would be an interesting method for analyzing histogenesis and that this should be easy to perform because it contains 'neoblasts' which are embryonic cells with migratory tendencies. In spite of this prediction, only about a dozen papers dealing with *in vitro* cultures of planarian fragments have been published. Some of them do not even describe true tissue cultures because wound healing suppresses the contact between parenchymal cells and medium so that the fragments behave as regenerates [C. SENGEL, 1960, 1963], or remain stationary isolates [MANELLI-NEGRI, 1962]. MURRAY [1927, 1928, 1931] was the first to attempt histotypical cultures and found a saline solution suitable for planarians. Addition of various amino acids showed arginine to be toxic. MURRAY made numerous observations, but her cultures did not live long enough and cells showed only amitoses. REISINGER [1959] mechanically dissociated the planarian pieces by forcing them through a gauze. Round aggregates were formed, covered with ciliated epithelium and containing a syncytium. SEILERN-ASPANG [1960a] cultivated minute explants on coverslips, in a liquid medium including placental serum. Migrating cells with clear cytoplasm were seeded in fresh medium and their shape changed after 14 days. SEILERN-ASPANG believed that these cells were 'neoblasts', capable of differentiation when isolated, but he never observed proliferation, either in explants or in isolated clumps of cells. ANSEVIN and BUCHSBAUM [1961] attempted to cultivate minute explants or clumps of cells on solid and in liquid media. Extracts of crushed planarian tissues were added as nutritive elements to a saline solution. Migration occurred on solid medium but only for a few hours. Isolated cells did not proliferate but died. More recently BETCHAKU [1967, 1970] cultivated minute prepharyngeal fragments in Holtfreter's solution or a slightly modified Eagles' medium in order to collect free 'neoblasts' and study their behavior. Explants were normally cultivated for 2 or 3 days and never tested beyond 7 days. Mitoses were observed in 'neoblasts' only. Gastrodermal cells near the cut surface were swollen with medium but no real growth was reported. BETCHAKU succeeded in isolating free basophilic cells whose life span has not been determined. There were no observations after a 48-hour period. Nevertheless, BETCHAKU concluded that cultured 'neoblasts' (inside or outside the explant) migrate, divide and aggregate as they are supposed to do in normal regeneration. This short insight of the literature in the field shows that histotypical cultures have never been obtained with planarian tissues.

More recently, FRANQUINET [1973] made up a new medium (including horse serum) that allowed tissue survival up to 12 days. Mitotic activity of 'regenerative cells' was observed to increase significantly. This result is extremely interesting especially when one realizes that one can think of the so-called regenerative cells as a sort of immunological defense system. By injecting intact worms with the same medium one could probably stimulate the mitotic activity in the same way.

We also attempted to establish cultures from *D. subtentaculata* and *D. gonocephala* [CHANDEBOIS, 1963a, 1968a] and later from *D. tigrina, D. lugubris* and *D. lacteum* tissues [unpubl. results]. The synthetic medium used is a mixture of MURRAY's salt solution and 13 amino acids (whose nature and concentration were determined by chromatography of planarian proteins). There was no arginine in the chromatogram and it was never added to the medium since as mentioned, it has been proved by MURRAY [1927] to be toxic. Extracts of crushed planarians were also added. To verify whether this composition was suitable for planarian culture, mitoses were counted in fragments placed in a nutritive medium and in controls of the same size, from the same level and put in fresh water for regeneration [CHANDEBOIS, 1968a]. The results obtained show that the culture medium clearly stimulates mitotic activity in type I cells, while in residual plasms and in the syncytium the rate of mitoses remains more or less unchanged (fig. 33).

Despite the above, just as in ANSEVIN and BUCHSBAUM's cultures [1961], groups of cells placed in the medium did not proliferate and finally died; very small fragments are rapidly dissociated, in spite of the solid support used. Some cells degenerate shortly after dissociation, others survive for several weeks. In order to prevent dissociation, it is necessary to culture relatively large fragments (approximately one tenth of the body length) in 1 ml of liquid medium, or on a solid medium including agar and covered with a drop of liquid medium. In this type of explant, the only type described by BETCHAKU [1967, 1970], the maintenance of co-

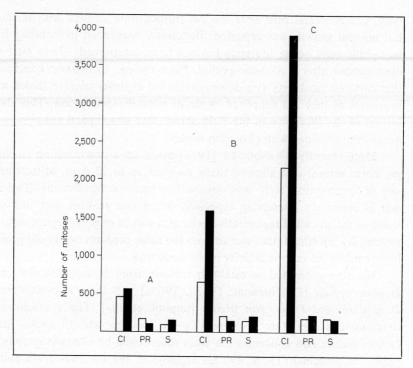

Fig. 33. Mitotic activity in cultured prepharyngeal fragments (black columns) compared with mitotic activity in regeneration of prepharyngeal fragments in tap water (clear columns). *A* After 2 days, fragments disaggregating *(B. subtentaculata). B* After 7 days, fragments with swollen gastrodermal cells *(D. subtentaculata). C* After 11 days, fragments with swollen gastrodermal cells *(D. gonocephala).* From CHANDEBOIS [1967].

hesion in the fragment is obviously related to a considerable swelling of gastrodermal cells that seal off the wound completely within 24–48 h (fig. 34A, B). After several days of culture, these fragments become very fragile. They disintegrate a few hours later if they are transfered to a fresh medium of culture or even if they are only pricked with a needle. Although in tissues of other metazoans dissociation stimulates mitosis (for which reason tissues are often artificially disaggregated before placing them in culture), in planarians it is never accompanied by visible proliferation. On the contrary, the fragments become smaller and smaller until the cells die. However, even in absolutely identical experimental conditions, not all explants behave in the same way. In some cultures, the gastrodermal cells do not swell, the epidermis rolls up, forming a kind of

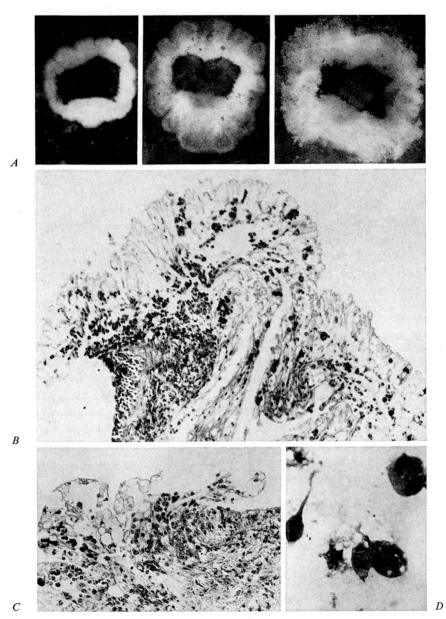

Fig. 34. In vitro culture of pieces of *D. lugubris*. *A* Morphology of the explants cultured for 1–7 days with swollen gastrodermis. *B* Histological section in a 7-day-old fragment. ×72. *C* Histological section in a 7-day-old fragment. ×72. *D* Type I cells with long pseudopodia collected at the surface of 7-day-old fragment. ×900.

belt around the fragment whose cells finally disperse. This behavior is observed in the explants of planarians which are starved or were kept in a laboratory for a long time. It occurs more frequently in *D. subtentaculata* than in *D. gonocephala*. In both species fragments taken from the posterior part of the prepharyngeal region keep better than the others. In certain explants, an epidermal wound healing takes place along the cut. If this healing process manages to envelop all the gastrodermal cells completely, a normal regeneration follows. This penomenon is fairly frequent in anterior prepharyngeal fragments, but only in the anterior cut surface; this is always more contracted than the posterior one and the mass of gastrodermal cells protruding through it is relatively small. On the other hand, explants placed on solid support will show cell dispersion. This will almost inevitably happen when the fragment is placed on agar without being covered by a drop of liquid medium.

The tendency of all sorts of explants to dissociate sooner or later and the subsequent degeneration of the dissociated cells seems at first incomprehensible. However, it becomes understandable on the basis of results obtained with electron microscopy (fig. 35). The isolation of a fragment necessarily leads to dedifferentiation and syncytium formation in that fragment. The remaining differentiated cells separate from each other because of the formation of the interstitial syncytium. The latter, since it contains practically no endoplasmic reticulum, flows free into the medium through any opening in the epidermis, so that the culture is finally reduced to isolated differentiated cells, incapable of dividing and forced to degenerate. If the gastrodermal cells swell with medium and seal off the opening created by the transection, they hold back the syncytium in the parenchyma. When the dedifferentiation of the fragment is extensive, a mere pin prick is sufficient to provoke its immediate disintegration.

When fragments do not disintegrate, they show no actual growth, regardless of the mitotic activity (which might be low, due to poor physiological conditions often created by viral infections, or completely inhibited by previous irradiation with X-rays). Cells outside of the fragments show no indication of proliferation. Only groups of gastrodermal cells of different numbers detach themselves. They take on a spherical shape, for they cannot adhere to the agar. They do not divide and, what is more paradoxical, the fragments which continue to show large numbers of mitoses start to decrease in size extremely rapidly, as early as the 2nd week in *D. gonocephala*. This decrease can partly be explained by the fact that type I cells (with extremely thin and elongated pseudopodia) tend to emi-

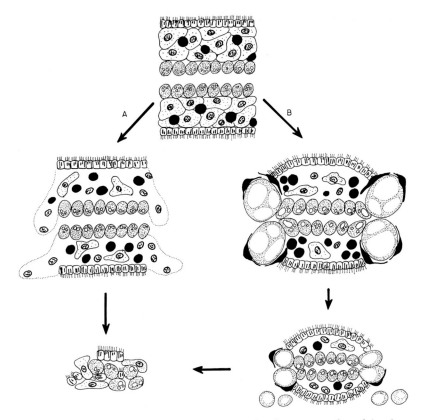

Fig. 35. Two possible transformations of planarian fragments cultured *in vitro*. *A* Parenchyma cells dedifferentiate; the resulting interstitial syncytium flows into the medium so that the cells of the fragment are disaggregated. *B* Syncytium is contained within the parenchyma because gastrodermal cells swell; type I cells migrate towards the medium, the size of the fragment diminishes, except when syncytium escapes, causing disaggregation.

grate from the explant (fig. 34C, D) in large numbers. This phenomenon corresponds to the migration of 'neoblasts' described by BETCHAKU. These cells can be gathered by placing a coverslip on the surface of the explant, for they are the only ones that can adhere to the glass. They are recognized by their intense basophily, by telophases that are not accompanied by cytokinesis, and by the production of buds that also look like very elongated pseudopodia. Considering the large number of cells and of mitoses that can be found, an envelope should be formed around the explant or layers of cells should appear on the surface of the agar, but observation shows that they are destroyed. Their phagocytosis was supposed to be accomplished by gastrodermal cells, but this has not been confirmed by histological examinations. It seems evident that cells of type I cytolyze in the medium in the same way as in the parenchyma. This hypothesis is supported by the fact that the number of mitoses in residual plasms in the explants is slightly inferior to the same in regenerating fragments in fresh water, despite the very high number of mitoses in type I cells. As mentioned, we succeeded in gathering on a coverslip type I cells that had migrated from a fragment and kept them alive in a hanging drop of medium. After 1 day, during which they emitted numerous cytoplasmic buds, they completely disappeared. The syncytium does not recover all the nucleic acids and is in a condition analogous to the conditions of starvation, i. e. a condition which could be the cause of the reduction of the explants. However, the decrease of fragments is much too rapid to be explained simply by an emigration and disintegration of type I cells. Observations of cultures seem to indicate that explants produce a greater quantity of gastrodermal cells. But this must be verified by accurate counts, because it is possible that absorption of culture medium by the gastroderm cells and their detachment from the epithelium produces the seeming increase in volume. Consequently, a definitive interpretation of these phenomenon cannot be proposed until one is familiar with the dynamics of planarian tissues and with the way in which this is modified in isolated fragments under the influence of a nutritive medium.

If one considers only the explants which do not dissociate during the first few days of culture and if one insists that there is no formal proof of cytolysis in the medium, one could conclude, as did BETCHAKU, that the classic 'neoblast' concept is correct. But if so, migratory and basophilic cells should be obtained in larger quantities when explants are taken from worms which are in the process of regeneration. However, in all such cases, it is the relative volume of the syncytium which increases. One-week-old regenerates can escape dissociation in culture, especially when one conserves a narrow band of tissue from the stump. After several days they may form large outgrowths consisting of syncytium that contains no free cells and whose mitotic activity is very high [CHANDEBOIS, 1963a]. Only a few type I cells are found in the parenchyma of the stump that is still attached to the explant. Fragments taken next to a regenerate, where morphallaxis is occurring, can escape dissociation thanks to the swelling of gastrodermal cells. In some of them, whitish outgrowths develop from accidental perforations in the epidermis. Histological examination showed that they are a kind of tumor, formed exclusively by the syncytium which contains numerous mitoses and is covered by an incompletely differentiated epidermis [CHANDEBOIS, 1968a]. Thus, in fragments taken from regions regenerating by epimorphosis or by morphallaxis, there are no more type I cells than in the others. There can be even less. It is the syncytium that is more abundant, an observation that proves once again that the classic 'neoblast' concept is insufficient to explain the facts. It is not impossible that the regeneration previous to the culture, by producing a massive cytolysis of type I cells, furnished a reserve of nucleic acids that the syncytium could use later on. This would explain why its capacity to proliferate is higher than the proliferation of explants from planarians which were not submitted to a previous regeneration.

In order to confirm these results, mitoses were counted on smears of explants, some taken from the anterior pharyngeal zone of normal planarians (fragments of type A), others from regenerating planarians, and representing either the regenerate itself (fragments of type B) or the adjacent zone of morphallaxis (fragments of type C). The material was from *D. subtentaculata*. The results were analyzed and compared with data from pre-

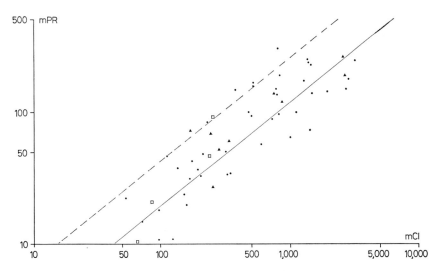

Fig. 36. Regression line of mPR against mCI for cultivated explants (●; *D. subtentaculata*). In addition to the graph, points given by cultivated regenerates (☐) and fragments from the morphallactic zone (▲) have been plotted. The regression line of mPR against mCI of experiment 1 has been drawn as control. From CHANDEBOIS [1972].

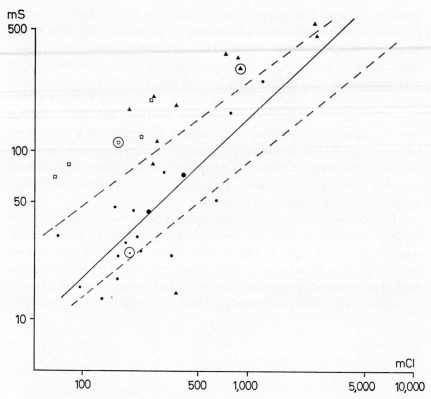

Fig. 37. Regression line of mS against mCI for cultivated explants *(D. subten-taculata).* Regression lines of experiments 1 and 2 (fig. 27) and points given by fragments A (·), B (□) and C (▲) as well as two heads which have regenerated tails for 9 days (●) are drawn. Points corresponding to arithmetical means are encircled. From CHANDEBOIS [1972].

pharyngeal fragments regenerating in fresh water [CHANDEBOIS, 1970, 1972]. The results are reproduced in figures 36 and 37. They lead to the same conclusions as the data discussed earlier (p. 59). The relation between mPR and mCI is modified by the medium of culture in exactly the same way for the three types of fragments. Once again this confirms that residual plasms and type I cells belong to the same system whose activity is not controlled by regenerative phenomena. In contrast, the relation between mS and mCI is not the same for all types of cultured fragments. It depends on the occurrence of a regeneration preceding the period of culture; this confirms the histogenetic function of the syncytium.

As was the case for irradiation, these experiments with tissue culture provide results that cannot be reconciled with the 'neoblast' concept but, in spite of many incertitudes, can be interpreted more logically with the processes of histogenesis as described in this study. This envisages the syncytium as an undifferentiated formative material and type I cells as actively synthetizing elements with an extremely short life span. Though the behavior of cultured fragments is at first difficult to understand, it finally appears as the logical consequence of the interstitial position of the syncytium and the practical incapacity of its nuclei to divide without the help of substances from type I cells.

Discussion

A wound in a planarian results in two reactions that can be dissociated from each other experimentally. On the one hand, as in any metazoan, there is dedifferentiation accompanied by activation of cells near the wound. On the other hand, in normally fed individuals there is a stimulation of a system comparable to the thymus of the vertebrates which is apparently found also in other kinds of worms. The resulting release of nucleic acid stimulates the proliferation of undifferentiated elements and thus minimizes the extent of dedifferentiation.

It seems now well established that in the simplest experimental conditions regeneration (epimorphosis and morphallaxis) involves two systems, whose activities can be dissociated experimentally from each other (fig. 38).

The first system provides totipotent material for the construction of the blastema and for tissue repair. It consists of an intercellular syncytium, that is a *transitory* and probably *short-term* phenomenon, between dedifferentiating and redifferentiating cells. It seems that adult, normally fed planarians do not have such syncytium in reserve. During regeneration dedifferentiation is restricted to stump tissues near the site of amputation. The syncytium which fills enlarged intercellular spaces is crowed with free ribosomes. Its mitotic activity ends when regeneration is completed, so that there is a close relation between the duration of mitotic activity and the amount of tissues that are to be reconstructed by the stump. Nuclear divisions provide most of the undifferentiated material and, consequently, limit dedifferentiation. Nevertheless, this is not always so, for the dedifferentiated cells may redifferentiate directly, without previous division.

The second system – the system of type I cells – has a trophic role that is related to the continual production of short-lived cells in non-

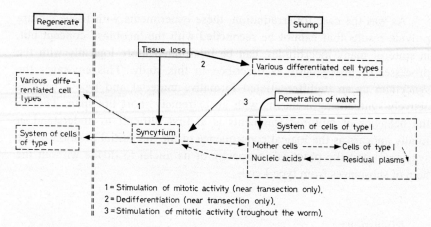

Fig. 38. The respective roles of the syncytium and of the system of type I cells in histogenesis and the factors which stimulate their mitotic activity during regeneration. From CHANDEBOIS [1972].

regenerating worms. Their exceptionally active synthesis of nucleic acids is demonstrated by the frequency of chromosome replication, by the strong basophily of the nucleoplasm and cytoplasm and by repeated amitoses and cytoplasmic budding. Another proof of this activity is the relative rapidity with which these cells incorporate labelled precursors of RNA, if compared with all other cell types in the worm. Cytolysis of type I cells and their cytoplasmic buds necessarily release all the products they have synthetized. Consequently, the type I cell system provides the syncytium with breakdown products of DNA that is indispensable for its proliferation. Since regeneration can take place without mitoses, this aid seems to limit dedifferentiation to a strict minimum. The type I cell system also provides the syncytium with RNA (or breakdown products of RNA) necessary for its differentiation, but should this be lacking the syncytium can produce a sufficient quantity of RNA on its own. In the present state of our research it is admittedly difficult to envisage how this material is utilized by the syncytium. It is highly improbable that any morphogenetic information comes directly either from DNA or RNA released by the type I cells.

The activity of the type I cell system is closely related to a nutritional factor. To recapitulate briefly what was already said before, it changes with the seasons and is progressively depressed until it stops during prolonged starvation. The system reacts to simple internal changes (such as penetration of fresh water via injections or incisions without tissue loss);

its mitotic activity rapidly increases and reaches a maximum 3 or 4 days later. The next few days, while mitotic activity decreases, the system is incapable of responding to another stimulation: it is 'cinephylectisized' [DUSTIN, 1923]. After amputation, water enters freely through the wound and stimulates mitotic activity throughout the parenchyma. For this reason the production of type I cells (and therefore nucleic acids) is not related to the amount of tissue that must later regenerate. It can thus be in excess, the surplus ready to be used for a while.

Certain experimental results obtained by GABRIEL and LE MOIGNE [1971a, b] and LE MOIGNE and GABRIEL [1971] (if reinterpreted with the above concept in mind) indicate that the type I cell system is particularly active during developmental growth. It has been already mentioned that in adult planarians one can obtain regeneration after RNA has been blocked by actinomycin D. But to ensure this, another incision has to be made beforehand that makes available a certain quantity of RNA [ZILLER, 1974a]. In the planarian which just hatches from its cocoon, this preliminary activation – very likely affecting the type I cell system – is not necessary.

This theory contradicts sharply the 'neoblast' concept widely accepted by biologists, especially since WOLFF's and DUBOIS' [1947] well-known experiments which gave the theory a firm foundation. Moreover, the new concept developed in this monograph and in the author's earlier work seems to complicate the solution of regeneration problems because it considers two systems which interact with each other in regeneration. Thus, the opposition to the author's view on the part of specialists is perfectly understandable. Even though a simple document such as figure 10 clearly shows that the 'neoblast' theory should be revised, the acceptance of the author's conclusions is difficult because of their novelty. Here is what BRØNDSTED [1969, p. 68] had to say: 'CHANDEBOIS' figures and interpretations are so unusual and involve so many completely new aspects of cellular activity that deep scepticism is called for. Only new investigations made by suitable techniques can clarify the problems completely, and until then I prefer to regard CHANDEBOIS' findings, interpretations and unproven hypotheses as due principally to erroneous techniques.'

This attitude is reasonable. If a theory would be proposed, applicable only to planarians and were to destroy current concepts of regeneration in all the other groups of metazoans, 'deep scepticism is called for' indeed. But in reality, if the syncytial and interstitial nature of undifferentiated tissue is put aside (probably related to aberrant embryonic development), the proposed outline does not contain any strange or unknown facts. If it is related to existing data in the literature – which will be done on the following pages – it becomes clear that planarians behave like all other metazoans ... and that it is in fact the 'neoblasts' concept that is unconventional.

Earlier the only possible explanation for planarian regeneration seemed to be the intervention of a stock of undifferentiated cells. In fact, at a time when one knew practically nothing about the nucleoplasmic relations nor about the interdependence of cells in carrying out their specific synthesis, differentiation was considered to be irreversible. NEEDHAM's [1942] well-known expression 'the doors are closed' reflects this uncompromising attitude. What is observed on histological sections was considered to be degeneration or a purely morphological loss of identity. As the 'neoblast' concept was generalized [WOLFF, 1961] the exclusive role of dedifferentiation in regeneration had to be proven in each group separately and accepted only after long discussions and much controversy.

In polychaetes, STEPHAN-DUBOIS [1958] attributed the role of totipotent 'neoblasts' to coelomocytes that invade the tissue surrounding the wound. The localized irradiation technique, as in planarians, seemed to confirm the interpretation of histological sections. But after much work, especially by CLARK and CLARK [1962] and THOUVENY [1967] dedifferentiation in each of the three layers is now recognized as the only means for recruting regenerative cells.

Similarly in arthropods where the healing of the wound mobilizes a great number of blood cells, it was thought that these may have a morphogenetic role in regeneration. But since the studies of HOARAU [1971] it is known that the hypodermis and the musculature regenerate on their own, and their cells pass through a temporary dedifferentiated state.

In amphibians, the problem of regeneration has a similar history. WOLFF and WEY-SCHUÉ [1952] tried to prove that the 'neoblast' concept was valid for this group too. LAZARD [1967] thought to have experimental proof because she revitalized irradiated limbs with the implantation of various healthy tissue. This evidence was refuted by DESSELLE [1968] who has shown that implantation of cartilage does in fact revitalize the irradiated limb, but it releases few cells. If the tissue is homogenized before it is placed in the irradiated limb, the result is the same. TRAMPUSCH and HARREBOMÉE [1965, 1969] demonstrated that dedifferentiation is a prerequisite of regeneration. It is not simply a temporary elimination of specific structures. By redifferentiating, the cells can provide another cell type with histogenetic potencies that seemed to have been lost after embryonic inductions in somites; the muscle redifferentiates into cartilage and inversely cartilage into muscles.

At the present time the concept that regeneration is due to the intervention of regenerative reserve cells is still held valid only for hydrozoans and for a few fresh-water oligochaetes.

In all cases where dedifferentiation has been recognized, histogenesis takes place in identical ways. The cells get rid of certain specific structures, either by autotomy or by autophagy. Consequently, lysosomes appear in cells near the wound and the activity of acid phosphatases intensifies. The endoplasmic reticulum disappears almost completely. The dedifferentiated cells concentrate on the synthesis of RNA and DNA. This phenomenon is called 'activation'. It may be recognized by histological examination which reveals an abundance of mitoses and the particular appearance of the interphasic cells: voluminous round nuclei, conspicuous nucleoli, scanty cytoplasm which is strongly basophilic.

In planarians it was almost inevitable that cells accumulating near incisions, in which PRENANT [1922] long ago recognized a striking similarity to the behavior of vertebrate lymphocytes, should be mistaken for regenerative cells. These exceptionally voluminous cells fill in the intercellular spaces where other, less conspicuous cells are dedifferentiating and merging to form an interstitial syncytium. Moreover, as type I cells take part in a particularly important synthesis of nucleic acids, they look much more active and morphogenetically potent than the dedifferentiating cells forming a syncytium. Nevertheless, activation in the syncytium is clearly discernible by electron microscopy. In the cytoplasm together with an abundance of ribosomes and polyribosomes there are also numerous lysosomes that have destroyed the endoplasmic reticulum. The appearance of the nuclei is also the same as in embryonic cells of other groups: rounded shape, clear and regular chromatin, large nucleoli.

The existence of a second system, linked to the digestive apparatus and activated by the penetration of water, does not seem to be peculiar to planarians. Their existence was reported in oligochaetes of the specific *Lumbriculus*. It was long believed that in these worms the big dissepimentary cells that are activated during regeneration are 'neoblasts' with mesodermal potencies. SAYLES [1931] studied the effects of freshwater injection. The injection needle injures the integument and the epidermis is activated only in the immediate vicinity of the wound. But the 'neoblasts' and the cells of the alimentary canal increase their mitotic activity in 10–12 segments near the wound. The worms react to the liquid, not to the needle itself, for Ringer injections have no effect. These observations are in harmony with data of STÉPHAN-DUBOIS [1954, 1956]. After amputation the epidermis which is an independent histogenetic system is activated only in the vicinity of the amputation, while the 'neoblasts' show mitoses in the next nine segments. Such an analogy with what one knows about planarians should lead to a reconsideration of the 'neoblasts' problem in oligochaetes. Research recently done in the author's laboratory by COULOMB-GAY and CORNEC [1973] with freshwater lumbricides *(Eiseniella tetraedra)* and leeches *(Herpobdella octoculata)* backs up this conclusion. After an amputation, acid phosphatase activity increases only in the immediate vicinity of the wound, in connection with dedifferentiation. But the alkaline phosphatase activity increases even far away from the level of the cut, but only in the endoderm. In this layer thus a particular system is stimulated by the incision, involved in rapid syntheses and its reaction is very different from cells that directly participate in forming the regenerate.

To attribute to type I cells a trophic, and immunological role would seem venturesome because no one has described anything of the sort in invertebrates. But the hypothesis is all the more probable since, in mammalians, it has been shown that nucleic acids released by the destruction of blood cells can be used by cells of regenerating organs. After injection of small lymphocytes labelled with tritiated uridine, labelled nuclei can be found in fibroblasts near healing wounds [MEDAWAR, 1957]. When hepatectomized mice receive blood whose leukocytes are labelled with ^{3}H-thymidine, labelling is later found in the nuclei of the regenerating liver [BRYANT, 1962]. There is an evident analogy between these studies and the experiments of LENDER and GABRIEL to which reference was made earlier on (p. 75). Moreover, several authors have reported that in amphibians intraperitoneal injections of RNA, from the same or from another species, stimulate regeneration in irradiated muscles [see references in WOLSKY, 1974].

If, as PRENANT [1922] has already shown, type I cells are analogous to circulating lymphocytes in vertebrates, they nonetheless seem to be closer to thymic lymphocytes. The vertebrate thymus has remained in the background and its functions have not been completely explained, despite a growing interest these last 10 years. However, by summarizing all that has been written or supposed about this organ, one finally establishes a mechanism that can be exactly identified with what was said above for the system of type I cells [CHANDEBOIS, 1972].

Thymic lymphocytes have a high nucleoplasmic ratio and a basophilic cytoplasm. They are continuously produced by mother cells. During their maturation, they exhibit high mitotic activity and they progressively diminish in size. DUSTIN and GRÉGOIRE [1931] came to the conclusion that chromosome numbers must be gradually reduced during abnormal mitoses. Few thymic lymphocytes are supposed to be found in the blood stream, but the majority of these cells are quickly destroyed *in situ* within 3 or 4 days. As there are only a few pycnotic nuclei in the thymus compared to the very large number of cells produced, the thymic lymphocytes are believed to have disintegrated. According to MILLER and OSOBA [1967] the significance of the production of these short-lived cells still remains a mystery; it cannot be linked to the immunological function of the thymus because the majority of thymic lymphocytes do not enter the blood stream. But it is important to recall that a trophic function of the thymus has been suspected repeatedly, although not yet demonstrated [METCALF, 1964]. Following lymphoid aplasia various diseases were observed, suggesting that thymic lymphocytes supply nucleic acids for growth and differentiation of other tissue cells [DUSTIN, 1920; ROMIEU and STAHL, 1949; LOUTIT, 1962]. DUSTIN [1923] observed that mitotic activity in cat's thymus is abruptly increased after an intraperitoneal injection of peptones. The number of mitoses reaches its maximum value within 3 or 4 days and then suddenly begins to decrease gradually. After such a stimulation, the thymus is 'cinephylectisized' that is it cannot respond to a second stimulation for a while. The thymus is particularly active during growth. The production of thymic lymphocytes is slowed down during fasting and enhanced when food is abundant. In poikilotherm mitotic activity in the thymus is lower in winter than in the summer [DUSTIN, 1912]. Thymus is much more sensitive to X-rays than other lymphoid systems [CRÉMIEU, 1912; MANDEL and RODESCH, 1966].

Today researchers focus their attention on the immunological function of the thymus. It is not known whether a similar function exists in the planarians but since BRØNDSTED's work [1969] it seems probable. This author associated two halves of planarians belonging to the related species *D. lugubris* and *D. polychroa*. The *polychroa* half is resorbed by *lugubris* which is slowly regenerated. BRØNDSTED thought that immunological processes are probably involved in the regression of *polychroa* tissues. It would be interesting to check whether type I cells are involved in these processes or not. It seems by no means impossible that the system of type I cells would be analogous to a primitive thymus scattered among parenchymatous elements, especially since thymus originates from the endodermal layer of the gill pouches of the vertebrate embryo and type I cells probably originate from the gastrodermis. If in two such different groups, phylogenetically speaking, as planarians and vertebrates, one finds two systems whose identical behavior suggests a trophic, immunological defensive role, it is certain that analogous systems must exist in other groups. Of course, these systems can remain unnoticed if the cells that constitute them are not so conspicuous, as they are in planarians, and do not form distinct organs, as in vertebrates.

In conclusion it may be said that there is good reason to believe that the type of cells which in the past, 'disguised' as embryonic cells, so often misled both observers and experimenters will be in the not too distant future generally recognized as having another role in the life of planarians, equally important as the role of embryonic cells and indirectly connected with regeneration.

Morphogenesis in Regenerating Planarians

The Current Concepts of Morphogenesis in Adult Planarians

The different concepts proposed as an explanation for the maintenance of the organization and the regeneration capacity refer to the idea of dominance: the head induces the other parts of the body and inhibits the formation of another head. These concepts cannot be applied to planarians where the various parts of the body have properties comparable to the properties of the head.

The dedifferentiated material used by the animal for forming the regenerate must necessarily receive a 'signal' from the stump to recreate the missing parts. Since the nature of this signal is unknown, the noncommital term 'induction' was applied to it. Yet this phenomenon is not comparable to embryonic induction for most, if not all the cell types differentiated in the 'induced' parts are also found in the 'inductor'. The specificity of the response lies in the spatial distribution of the cells, that is, the 'inductor' essentially determines the patterns of the 'induced' region. Thus, the term 'morphogenetic induction' used by ABELOOS [1932] is preferable.

Studies in WOLFF's laboratory have shown that 'induction' does not result in the simultaneous formation of all the organs. 'Induction' is the starting point of *epigenetic phenomena* whose first consequence is the differentiation and organization of the blastema, then the achievement of morphallaxis. According to LENDER [1952] who did experiments with *P. nigra,* the brain exerts an induction from a distance on the edge of the head (this is a real embryonic induction). Thus, the eyes appear here, even before the nerve fibers have regenerated. This 'induction' takes place via substances that are found in the supernatant of homogenates of cephalic centrifuged tissues. In the parts that regenerate by morphallaxis, epigenetic phenomena have also been discovered. P. SENGEL [1951, 1953] showed that for the formation of a pharynx first a pharyngeal zone has to be established. Regeneration appears as the *performance of a program* by the dedifferentiated material, but the exact stages and mechanisms still have to be clarified.

The stump tissues are evidently in charge of programming. A head blastema, a tail blastema and a lateral blastema do not have the same program. How does the stump choose, for each one of them, an adequate program so that the reformed parts will reproduce the missing parts? These questions that we express in contemporary terms were raised a long time ago for planarians, as for animals and plants in which regeneration can take place in two or three dimensions.

With the help of regeneration various parts of the body reveal their capacity to reconstruct the others. One can say that regeneration manifests the 'morphogenetic potencies' of the body which are not used in the uninjured animal. Implicitly this idea of morphogenetic potential means that each region, considered as a whole, must choose, at any given moment, among several types of organization and that the final choice can only be made by factors that are *extrinsic* to it. That is why the stability of the adult organism was attributed for a long time to correlations among its various constituent regions. Therefore, problems posed by the maintenance of the unity of organization were naturally associated with problems of regeneration.

Current concepts dealing with adult morphogenesis are still greatly influenced by the theory of *metabolic axial gradients* proposed by the American embryologist CHILD [1941]. This theory, equally applicable to embryogenesis, regeneration and asexual reproduction, is based upon the existence of continuous variations in metabolic activity along one or more axes of polarity. These variations were shown experimentally in different groups, either by measurement of oxygen consumption, or by decoloration of individuals stained with vital dyes under anaerobic conditions, or by differential disintegration brought about by cyanide. The gradient is probably maintained by the correlations between the various levels; each of them slows down the metabolic activity of those that are lower in the gradient. For CHILD, to each quantitative degree of protoplasmic activity there is a certain corresponding quality of morphological and histological differentiation. Thus, a given level owes its morphological pattern to the physiological activity of the level that immediately precedes it and, in turn, organizes the next level. Its dominance is *relative* since it is both dominant and dominated. *Absolute dominance* belongs to the 'high point' of the gradient, that is the region where the metabolism is most active. Considering that in the uninjured animal the morphogenetic capacity is not expressed, dominance is not only a kind of morphogenetic induction, but also an inhibiting action which spreads along the gradient and maintains the unity of the individual's organization.

In a planarian the head is the high point of the metabolic axial gradient. So, according to CHILD, it has absolute dominance. The head's organizing action seemed well demonstrated by the grafting experiments of SANTOS [1929, 1931] reproduced by many authors. A head grafted in the postpharyngeal region of a whole planarian completes itself with a prepharyngeal region, then with one, or two pharyngeal regions (fig. 39B). The idea that the worm's head prevents the formation of other heads and maintains the unity of organization seems to be proved by CHILD's [1941] experiments in which a sagittal notch separated the right and the left halves of the anterior part of the body (fig. 39A). In each of them, the cephalic end was removed at different levels. If the more posterior of the cuts is at the bottom of the sagittal notch, the blastema is linked to the other half (where a head is rebuilding), and its development is inhibited. If this cut is made at a more anterior level, the two blastemas are separated from each other, and each develops into a complete head, for inhibition no longer has any effect.

It soon became evident that simple quantitative variations of metabolic activity could not explain alone all the phenomena of organization. Thus, other concepts were proposed, but they all retained the main idea of CHILD's theory: absolute unilateral dominance of the cephalic region over the other parts of the body.

At present time most specialists of planarian regeneration accept the concept formulated by WOLFF, from 1953 to 1964, with the collaboration of SENGEL, LENDER and ZILLER-SENGEL. Dominance seems to be explained more precisely because it is attributed to the activity of 'inductive' and 'inhibitory' substances. These substances are specific to the region or the organs that produce them. Their concentration is highest where they are synthetized and decreases gradually as they spread by diffusion; this would explain the existence of axial gradients. Moreover, the action of the morphogenetic substances is competitive at each level. WOLFF [1953] first believed that regeneration is a succession of 'zone inductions' that trigger each other; he described them as they would occur in the regeneration of postpharyngeal fragments. First a blastema forms on the site of amputation and becomes a head, as determined by the stump. A brain, comparable to an 'organizing center' in amphibian embryogenesis is differentiated; it induces the eyes. The epimorphic regenerate induces a prepharyngeal

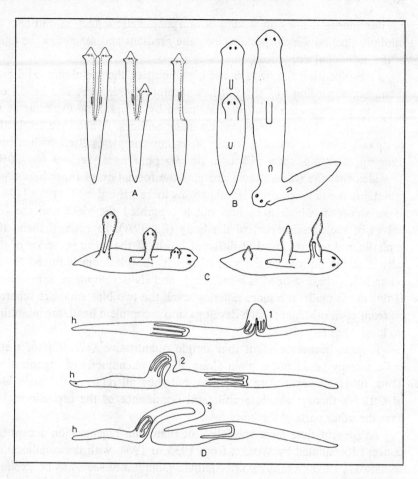

Fig. 39. Experiments dealing with dominance. *A* [CHILD, 1941], *B* [OKADA and SUGINO, 1937], Experiments which support the theory of head dominance. *C* [SCHEWTSCHENKO, 1937], *D* [OKADA and SUGINO, 1937] Experiments which prove that any region may be dominant upon the others. 1 = Prepharyngeal piece into postpharyngeal zone; 2 = midpharyngeal into prepharyngeal zone; 3 = postpharyngeal into prepharyngeal; h = cephalic end (head).

zone in the stump tissue; the latter induces a pharyngeal zone whose lateral parts induce a pharynx. Finally, the gonads of this zone induce the copulatory organs. According to WOLFF, these inductions can only take place in the *head-tail* direction. In the case of posterior regeneration, the reconstruction of the pharynx is determined by inductions coming from the head and the prepharyngeal zone of the stump. The epimorphic regenerate can only produce a postpharyngeal zone. Later on, WOLFF [1962], then WOLFF et al. [1962, 1964] ascribed a more and more important role to inhibitions. Thus, the brain releases a substance that inhibits the differentiation of another brain and spreads right to the posterior extremity. Likewise, the prepharyngeal zone inhibits the formation of another prepharyngeal zone and the pharynx the formation of another pharynx in the more posterior zone. The experimental verifications are based on simple methods. The tissues of a zone or of an organ are homogenized and centrifuged. The collected supernatant is added to the water in which the animals whose amputated part provided the extract, are starting to regenerate. This technique was first used for the brain by LENDER [1955, 1956, 1960]. All the planarians regenerated a head and eyes (because of the 'organizines' present in the homogenates – according to the author), but in 20 % of the cases no brain was reconstituted while in the others it was very small. With crushed tails total inhibition never happened and the average size of the brain is bigger. ZILLER-SENGEL [1965, 1967a–c] did the same type of experiments for the pharynx. Here regeneration is never inhibited but only delayed. The delay is shorter with head homogenates. ROSE [1970] related analogous phenomena in different groups, especially the hydrozoans and the annelids. He allots the unity of the individuals' organization to inhibitors whose elimination sets off regeneration.

This concept seemed further strengthened by the fact that induction and inhibition of one region of the organism by another are well-known phenomena for the embryo, both on the cellular and molecular level. It has been proven that a single substance from a primordium on its own determined another primordium (e. g. RNA protein taken from the notochord induces cartilage in the competent somites [LASH, 1963]). There have also been precise cases of inhibition that prevented the manifestation of certain histogenetic potencies (e. g. the inhibition of the differentiation of a Wolffian lens regenerate by grafting epidermic lens [STONE, 1958]). However, this concept, like CHILD's orginal concept, leaves a number of questions of planarian regeneration unanswered.

WOLFF [1962] analyzed in detail the case of anterior regeneration, but did not elaborate on posterior or lateral regeneration. 'A regeneration blastema can be considered as a graft of undifferentiated elements on the adult organism ... It can be assumed that the base induces the formation of an *anterior*[4] blastema, in which the first differentiation is that of a brain. The brain has an inducing action, through the intermediary of diffusible substances, on the eyes.' This mechanism does not explain, how a planarian can produce three types of blastemas: anterior, posterior and lateral. If their determination were simply due to induction, the choice among these three types could be interpreted in two ways: either there are various substances that respectively induce heads, tails or lateral parts and circulate in different directions [BONNET supported this theory in 1745] or the blastemas' diversity comes from the diversity of undifferentiated cells that make them up, an assumption that is equally unreasonable. However, the brain could really represent an organizing center and in this case the question is what would be the organizing center in a tail blastema or in a lateral regenerate? These questions still await answer.

The idea that each zone is induced by a more anterior zone and in turn induces a more posterior zone (in anterior regeneration as well as in posterior regeneration) is based on incomplete experimental data. Indeed a head or a prepharyngeal region implanted in a postpharyngeal region can induce a pharynx there [SCHEWTSCHENKO, 1937; OKADA and SUGINO, 1937]. But it is usually forgotten that the same authors who reported this also described the induction of a pharynx in the prepharyngeal region of a planarian after a postpharyngeal piece had been grafted there (fig. 39C, D). In the light of these results it is difficult to explain why the head grafted in the prepharyngeal region did not induce a pharynx [SCHEWTSCHENKO, 1937; TESHIROGI, 1956]. Within the framework of the current concept, it is logical to think that the action of an inductive substance, released by the graft, should be reinforced rather than neutralized by the action sent out by the host head. In every case where a graft produces a pharyngeal zone by induction, the polarity of the induced structure conforms to that of the graft. This is the case even if the structure is formed by morphallaxis in the host tissue. Consequently, the polarity of the induced pharynx is often inverted. As it is already very difficult to explain how the diffusion of an inductive substance (or the elimination of an inhibitory substance) can by itself convert a pre- or postpharyngeal region into a pharyngeal region, it is even harder to understand how it could also reverse its anteroposterior polarity.

Among the grafts made by OKADA and SUGINO [1937] there are a number of cases which do no support the 'induction by zones' idea as

4   Italics mine.

defined by WOLFF. Particulary significant is in this connection one experiment (fig. 41 IV). The cephalic and caudal ends of a worm are amputated. The rest is cut in two by a transversal section at a level slightly anterior to the pharynx. The anterior cut surface of the anterior piece is joined to the posterior cut surface of the posterior piece. At the level of the junction, a supplementary pharyngeal zone appears, *whose polarity is inversed in comparison with the polarity of the two joined fragments.* Obviously the region is integrated neither with the one nor with the other of the two pieces which are joined: none of the zones completed itself at the expense of the other. The only possible explanation was given by the authors: 'The formation of the new tissue implies a lack in a physiological or morphological sense between the *cut surfaces*[5] which have been brought into union.' In other words, neither the anterior nor the posterior fragment dominates; the intercalary regenerate restores structures which, in the normal organism, are lying between the two cut surfaces, experimentally brought together. The poor choice of word (lack) obviously means some sort of 'vacuum', or gap.

The hypothesis of the diffusion of inhibiting substances preventing the reproduction of already existing structures is equally without a solid basis. The duplication of an organ is very frequent in planarians. The various types of abnormal regeneration which can be attributed to physiological disturbances will not be discussed here, but those produced on purpose by grafts and incisions need some comment. The supernumerary organ is not always very far away from its counterpart that is in a region where the inhibiting substance that reaches it would not be too diluted to have an effect. There are numerous ways of making a second head appear near the stump's head. Similarly, a supernumerary pharynx can appear next to another after various types of grafts. One cannot assume that in such cases 'dominance' of the graft is superior to the pharynx's inhibitory capacity since any grafted region (cephalic, prepharyngeal or postpharyngeal) can bring about this induction.

Another weak point of the dominance and inhibition concept is shown by the spontaneous abnormalities that occur during blastema determination (heteromorphosis and hemiheteromorphosis) and the anarchy of regeneration such as the ones shown in figure 4.

It is interresting to note that the concept of head dominance was seriously questioned by BRØNDSTED [1939] after experiments with *D. lu-*

---

5   Italics mine.

gubris. Short prepharyngeal fragments were isolated by two transverse cuts and a head was immediately grafted on the posterior cut surface. On the anterior surface a head regenerate with normal polarity appeared and it differentiated eyes at a normal rate. When this regenerate was removed, it was replaced normally again. Heads were also grafted on the anterior wound surface of prepharyngeal fragments, but with dorsoventral polarity inverted. Normal head regenerates still formed at the level of this cut. In the first experiment, the grafted head was not able to reorganize the tissues of the prepharyngeal fragment; it did not seem to invert the tissues' anteroposterior polarity, even after several days. It could not inhibit the formation of another head at a relatively short distance. The second experiment is much more conclusive; the cephalic regenerate of the fragment developed in a normal way in immediate contact with the grafted head. As for the experiments that seemed to demonstrate the existence of inhibiting substances, their results are not decisive. In fact, these substances (which lower regeneration but do not always completely block it) are obtained by crushing the tissue and the homogenate artificially brought into contact with the regenerating fragment. Nothing in the present state of research can prove that these inhibiting substances penetrate the intercellular spaces and spread throughout the uninjured organism. They are perhaps released by broken cells only and may be even modified by hydrolases which are activated by crushing.

During the last few years, some biotheoreticians [WOLPERT, 1969, 1971; GOODWIN and COHEN, 1969] have justly criticized the current tendency to search for clues to morphogenesis at the molecular level. They particularly emphasized that the importance given to inducing and inhibiting substances almost completely overshadows the problem of pattern formation. Each substance comes from cell activity, conditioned by intercellular relations. Thus, the concept offered by these authors is based on a new notion *'positional information'*, that is, information that allots to each cell its place in the system at the same time that it determines the nature of its specific metabolic activities. Several 'reference points' can be found in a system; the differentiation of each cell depends upon its relative position with respect to these different points. For example, in a system with an axis of symmetry, the two poles of the axis are the reference points. At their level, certain properties (for example concentration of a substance) have different values and between the two of them, these same properties generally vary. WOLPERT supposes for example that at one end of the axis (the 'source') the cells release a substance; the cells of the other end (the 'sink'), make the substance disappear. Between the two, the concentration of this substance varies according to a gradient and with this variation goes the value of positional information. After an amputation, the edge of the cut will regain the boundary value of the eliminated part, thus restoring gradient of the remaining part of the system. Consequently, this part builds the structures of the initial system, with reduced cell material. WOLPERT's concept starts with original and very constructive remarks. It applies perfectly to certain embryonic systems. However, the idea that the transport of substances over long distances fixes specific positional information at each level does not apply to adult planarians. In fact, in any species, one often obtains little fragments, at any level, that heal but do not regenerate. If one disregards the dedifferentiation that takes place on the cut surface, these fragments remain exactly what they are. A prepharyngeal fragment remains a prepharyngeal fragment just as a pharyngeal fragment remains pharyngeal. A head does not develop a pharynx. If the operation separates a region from a positional information orginating from other regions (e. g. head and tail), they will undergo morphallaxis which converts them as *a whole* into another region. *Thus it is certain that any zone of the body, even a very small one, can keep its structures intact without the intervention of the rest; correlation among its own constituent parts necessarily assures this stability.*

At present, there is no concept that takes into account all the experimental data provided by planarians. The shortcomings of the theories are a perfect illustration of a common fate, of which biologists are becoming more and more aware: theories are appealing as long as they are not confronted with all the available relevant data. A single forgotten fact, if remembered in time, can sometimes destroy a whole concept. On the other hand, even the most careful survey of all the facts that we know about regeneration is of little use without a basic overall concept. Otherwise, the facts themselves remain a catalog of scattered data and it would be impossible to propose new experimental approaches. But one must not be content to amend old theories; occasionally it is necessary to blaze completely new trails.

For technical reasons, it is impossible to perform on regenerates the same type of microsurgical operations which were so successful with the analysis of epigenetic processes during embryonic development for example in sea urchins or amphibians. Because of this handicap, the experimenter, like the theoretician, naturally disregards structure or extremely simplifies it. Some biologists thus put themselves above all organization.

With this attitude, one can only obtain the type of concept for which words are more important than facts and theories more important than certainties. The opposite tendency, due to the success of molecular biology, works on a lower level, with concepts of metabolism. The experimental program is technically feasible and practically inexhaustible, but we share WOLPERT's pessimism. Even if we will possess an inventory of all the substances whose concentration or activity varies during regeneration, the fundamental problems will be still not solved. Between the two levels, the organological and molecular, there are the organism's anatomical and physiological units: the cells. Their properties give stability to the tissue in normal conditions and the plasticity that they show in experimental conditions. The substances are produced by the cells' cooperation with other cells. In other words, morphogenesis is, above all, *cell sociology*[6]. It is the integration of all the particular facts of cell associations which enable us to establish a descriptive model for the total phenomenon.

The Integration of Cell Activities in the Adult Organism

The primary integration system is considered to be a cell transformation system in which all the differentiated types can be converted into each other. Each level along the axis of the body is specified by a particular equilibrium between cell types, maintained by the adjacent levels' equilibrium. When a level is removed from its neighbors, certain elements dedifferentiate in the remaining part. To redifferentiate, and reestablish equilibrium of eliminated parts, the latter must receive new information that can only come from contact with a distant level of the system, which is normally not adjacent to it.

Up until now biotheoreticians have believed that adult regeneration and embryonic regulation are the same kind of phenomena. This point of view is justified if only the final result is considered which is the same: repairing of deficiencies. Yet the fact that morphogenesis does not take place in the same way in the two cases has not been emphasized sufficiently [CHANDEBOIS, 1973b]. During regeneration the blastema is sharply different from the stump and the following morphallaxis consists of a metamorphosis of tissues in the vicinity of the blastema. Nothing of the sort happens in the embryo. No blastemas are formed and the regulations observed have nothing in common with morphallaxis. Deficiency occurs

6   This concept is developed in a general treatise: R. CHANDEBOIS, Morphogénétique des animaux pluricellulaires (Maloine, Paris, 1976).

when the primordium is still undifferentiated. The repair is observed only in the later stages when the embryo is organized. From the point of view of cellular sociology, the difference between embryonic regulation and adult regeneration is obvious. In the embryo all the cells are differentiating. While they repair a deficiency, they continue to differentiate, but in a changed direction, in the one which the eliminated cells would have chosen. In the adult the tissues have reached definitive differentiation. When they repair a deficiency, they first have to make a step backward, i. e. dedifferentiate. The properties of embryonic organisms are only found temporarily in the regenerate where a partial ontogenesis is taking place.

Cellular correlations are infinitely more varied in the adult than in the young embryo and much less is known about them. Before trying to find out how the blastema receives, and carries out, the regeneration program, one must first understand exactly how the cell activities are integrated in the intact adult, so as to insure the stability of its structures.

We know that adult differentiated cells, withdrawn from their normal environment (isolated in a culture or transplanted) change their metabolic activities. This is manifested by an increase – or decrease – in their mitotic rate and especially by the stopping or modification of specific syntheses. It has been proved numerous times that in developing organisms cells that are topographically close to each other transmit mutual information discharging into the intercellular spaces substances which they have produced (in some particular cases, ribonucleoproteins). This is equally true for the adult. Because of this interdependence the cells of the organism are integrated systems in which each cell uses only part of its capacities for synthesis and multiplies at a rate imposed upon it by the system as a whole.

Despite their static appearance, these systems in the adult have proved to constitute an extremely complex machinery. Their function *at a given time* cannot be reduced to the diffusion of substances in the present. Let us first suppose, for the sake of simplifying the problem, that the individual which lacks a circulatory system is made up of reversibly differentiated cells whose life span is the same as that of the organism itself. Each cell has an *individuality* (metabolic rate, specific metabolism, adhesiveness) with respect to its position in the whole. *It has an endowment from its cell lineage, i. e. all the extracellular information received during development,* in the course of successive inductions. To preserve this heritage, that is to maintain its specific activities, each cell must have a particular environment. Essentially this is provided by the substances released by the neighboring cells into the intercellular spaces. The neighboring cells have also their special heritage and can be dedifferentiated if they are isolated. Thus it is obvious that *cellular interactions in the adult do not create the individuality of the cell; they just maintain it.* As each cell helps its neighbors to remain in a differentiated state, the organization of the whole is maintained. This fundamental phenomenon can be illustrated by comparing the animal to a 'semicircular arch' in architecture: no one stone holds up the whole; once they are cut and put into place, the elements hold up each other collectively. However, cellular equilibrium in the organism is not as simple as a piece of architecture because it is not static. Most differentiated cells are rapidly used up; they are destroyed or eliminated in large quantities. The organism continually repairs deficiencies, replaces the worn-out cells and reestablishes the constantly upset equilibrium. Every cell that replaces another takes over also its individuality, thanks to the information received from its environment. It is also known that this renewal may sometimes involve extensive migrations [BURNETT, 1967]. In *Hydra* for example, epidermal cells are produced at the hypostome and destroyed both at the tips of the tentacles and in the foot. Likewise in certain tubular glands, mitoses are observed at the base, dying cells at the opening. In these cases too, the position of the generating cells, the rate of their divisions, the differentiation of the cells they produce are determined by surrounding elements. Reciprocally, cells with a short life span following a certain passage, release substances and thus maintain the metabolic activities of cells they pass on their way. To conclude, one can say that the individuality of the cells that are present – or succeed each other – at any point of the organism, is maintained by, and maintains in its turn, the individuality of topographically related cells. Theoretically these correlations are sufficient for the maintenance of patterns in the adult. They are put in place during ontogenesis. They represent the principle of the *primary integration system.*

In coelomates, the start of blood circulation at the beginning of the functional phase of organogenesis establishes a new system for integrating cellular activities: a *secondary integration system.* In fact certain secretions especially hormones, are spread out throughout the organism. The cells can react everywhere by changing their metabolic activity. But what is more important and is well-known to endocrinologists is that *each cell reacts in accordance to its individuality, maintained by its integration into the primary system.* Consequently, substances spread throughout the system can modify to some extent the organization of various regions of the individual, but they obviously cannot obliterate the primary integration system. This reminder from elementary biology is indispensable for showing that the hypothesis of a release of inhibitory or inductive substances by any part of the body cannot fully explain the maintenance of the organization. By imagining a polarized diffusion, with a progressive lowering of the concentration along the axis – as in WOLPERT's 'source-sink' system – the most one can hope to obtain is the maintenance of a gradient; but the metazoans' patterns are much more complicated. *A fortiori* one does not understand how simple quantitative modifications following an amputation could lead to the transformation of one part of the primary integration system into another one, without which the genesis of new structures during regeneration is impossible. It is certain that the secondary integration system takes part in regeneration (for example amphidynamic earthworms regenerate during diapause only their tail). But one must emphasize the fact that certain hormonal interactions play a role in the development of the regenerate. Theoretically they cannot intervene in its determination.

The essential thing in a regeneration is that material, which belonged to the structures of a certain region, is rapidly converted to form a new region. Not all the cells have to return to a completely undifferentiated state, as shown in planarians by the persistence of the epidermis during blastema formation and, better still, by morphallaxis. In the course of regeneration new tissue types may appear. The ways in which this modulation takes place are well known. The cells of tissue X can modify their specific metabolic activities when a new extracellular factor reaches them. They change directly into a new type Y. This is *cell modulation*. More frequently, between the two successive differentiated states X and Y there is a period where the cells temporarily lose the ability to produce specific substances. Their specific structures disappear and they return to an undifferentiated state. After a short period of activation, these cells redifferentiate, possibly into Y if in the meantime their environment was modified and provided adequate information. When type X cells are renewed, the differentiation is sometimes irreversible. However, they can be replaced by type Y cells, if the generating cells have kept the corresponding potencies and if the daughter cells receive adequate information at the time they enter the differentiation phase. This is the mechanism of *tissue modulation;* it does not imply a change of specific metabolisms in the differentiated cells.

For a region to be able to transform itself into another region, the fundamental condition is that this region includes cells which are able to generate all the cell types of the other. With this as a basis, *a cell transformation system* was defined as one in which all the differentiated cell types are interconvertible [CHANDEBOIS, 1963a, 1965a]. This implies that the system has only one type of undifferentiated cells which can produce all the differentiated types of the system. These can be either cells which were always undifferentiated and present in the organism as such or cells in a transitory state between two successive differentiated states. The latter type is found in planarians. In fact, in the unique transformation system of these animals, the undifferentiated components (i. e. the totipotent interstitial syncytium) are only present in the case of a massive dedifferentiation. Since experimental data are lacking, one cannot define precisely the dynamics of the various tissue categories that belong to this system. It seems that the epidermal cells continuously replaced [SKAER, 1965]. During regeneration, its dedifferentiation does not completely obliterate its distinctive traits. Muscular cells seem to have a longer life span but can return to a completely undifferentiated state.

Some observations suggest that the essential result of cellular interactions in the transformation system is the maintenance of a numerical balance between the various categories of differentiated cells. Otherwise one could not explain that planarians submitted to prolonged starvation reduce harmoniously, although the number of their cells decreases continuously [ABELOOS and LECAMP, 1929]. The abundance of syncytium in starved planarians offers a possible interpretation of the phenomenon. It is possible that in tissues undergoing renewal mitoses do not suffice for compensating the losses. Thus the number of their cells diminishes. In such cases they release a smaller amount of substances necessary for maintaining the cells of other tissues of the organism in their differentiated state. Consequently, some cells of these tissues will dedifferentiate.

Since a cell transformation system is organized, there is a characteristic balance in each of the regions of the body along the main axis. For purposes of clarification, let us imagine a region of the organism, very small in size, arbitrarily defined. A simple histological examination shows that it is determined specifically by the nature, proportions and spatial distribution of the cell types that constitute it. This region is obviously characterized by a particular total metabolism (that can be referred to as 'metabolic level') which is indirectly linked to the rate of tissue renewal. The tissues which make up this arbitrarily delineated region are necessarily influenced by tissues of the neighboring regions. This is the basis of the state that the tissue equilibrium of any part of the system is maintained by

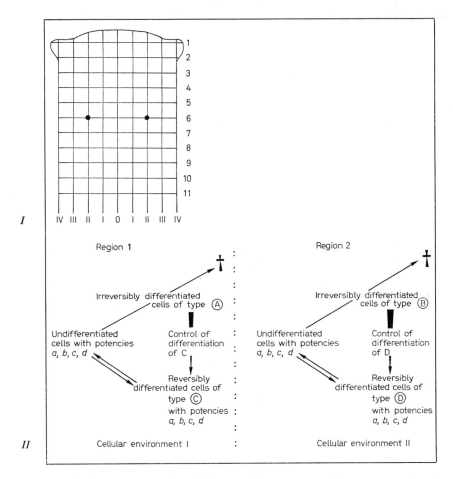

Fig. 40. Integration of cell activities in adult planarians. *I* System of graphs. *II* Mechanism of conversion of one region into another as explained by the concept of the cell transformation systems. The cellular environment of the region 1 being changed, undifferentiated cells produce B instead of A. Consequently, the type C is replaced by D (through modulation or dedifferentiation). From CHANDEBOIS [1973b].

the tissue equilibrium of the adjacent parts. Should this equilibrium in the adjacent parts change, the tissues in the region under consideration can show modulation, the rate of tissue renewal can change, the tissues can modify their numerical balance (fig. 41). Because of the integration of the various parts of the system, the alterations progressively spread, until a new state of balance is reached by the entire system. However, this restoration of a new balance implies that none of the differentiated cells will resist the propagation of transformations provoked by a localized imbalance. Dominance intervenes here. It is known that certain differentiated cell types cannot respond to modifications of their environment by a change in specific activity. However, they produce and release into the intercellular spaces substances to which other cells are sensitive. It may be assumed that these cells with unchangeable metabolism will impose their specific activities on other types. At the same time, they exert an inhibitory influence on some of the potencies of other cell types. These are the properties which CHILD envisaged as absolute dominance, but transposed here from the gross morphological to the cellular level. In most cases the differentiation of cells is reversible and in planarians this is always the case. The tissues control the activity of neighboring tissues and are in turn controlled by them. The dominance is relative and the different categories of cells and tissues necessarily have different degrees of dominance. The various differentiated types are not confined to a particular area, but are spread out throughout the organism. Consequently any part, even a small one, both dominates, and is dominated by, other regions. For example, in planarians, the 'induction' of a supplementary pharynx by a graft is considered to be the most typical case of dominance. Experiments prove that the postpharyngeal region is dominated by a head or by a prepharyngeal fragment that is grafted onto it, but conversely a postpharyngeal region dominates the prepharyngeal region when transplanted at its level (fig. 39C, D).

Through purely theoretical considerations, one can reach the same conclusions as the author did on the basis of her studies on the marine planarian *Procerodes lobata* [CHANDEBOIS, 1957a]. It was stipulated in that study that a planarian body is made up of an almost infinite number of regions, perhaps infinitesimally small, each having particular metabolic properties and exerting a reciprocal dominance on each other. It is now suggested that each of these regions is defined by a particular tissue equilibrium that is maintained by cells of neighboring regions. All these regions together make up the *primary integration system*. It seems that the diagram that was proposed in 1957 is still valid. Regulation phenomena are manifested in the two morphological dimensions of the body: longitudinal and transversal. All the regions situated on the same line perpendicular to the anteroposterior axis constitute *a particular level of the anteroposterior system*. Similarly, the regions situated on the same line parallel to the anteroposterior axis make up *a particular level of the me-*

diolateral system. Every level is characterized by the type of equilibrium of its constituent regions, which are in turn specified by the types of equilibrium of the adjacent levels. The system of graphs used in 1957 for this concept is reproduced in fig. 40. A certain number of the levels in the uninjured animal are represented by straight, parallel, equidistant lines. The levels of the transversal system are numbered with Roman numerals, starting in the median line and increasing in right and left direction. The levels of the anteroposterior system are numbered with Arabic numerals, number 1 representing the anterior edge of the head. There should be also a dorsoventral integration system, but its activity in planarian regeneration is negligible.

Regeneration consists in the conversion of levels of one system into others. Because the various parts of the organism are well integrated, the conversion can only take place if a level is separated from its immediate neighbor and then placed in the vicinity of another level in the system. The respective equilibrium of these two levels put side by side is necessarily upset; consequently certain tissues dedifferentiate. But at both levels, the dedifferentiated cells receive new extracellular information through contact with the other: they redifferentiate, setting up a new balance. Let us suppose, for clarity's sake, that levels 2 and 10 are placed side by side. The dedifferentiated material of 2, when it comes in contact with 10, receives new information. While redifferentiating it can but set up the balance of the level *which is close to 10 in the normal system,* that is level 9 (one can also imagine that if 10 dedifferentiated more completely or more quickly, it would be reconverted into 3 by level 2). Since the imbalance persists, levels 8 (or 4) are formed, and so on, until a complete sequence in the levels is reestablished. This phenomenon manifests itself in *intercalary regeneration,* whose sole origin is the interaction of two levels put side by side, *without intervention of the other levels of the connected parts.*

This theory is confirmed by the remarkable experimental work done by OKADA and SUGINO [1937]. These authors perfected a grafting technique that permitted them to recombine systematically three regions of the body: artepharyngeal (*a. p*), pharyngeal (*p*) and postpharyngeal (*p. p*).

To facilitate the presentation of their results, in terms of the schema described above let us give arbitrary number of levels to the anteroposterior system: 1 and 2 for the head, 3–6 for *a. p,* 7–10 for *p* and 11 and 12 for *p. p.* OKADA and SUGINO worked out four types of combinations (fig. 41).

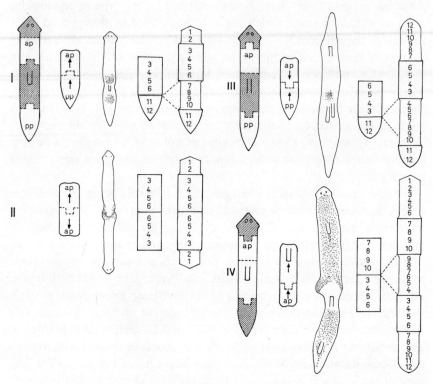

Fig. 41. Intercalary regeneration. Explanation in the text. From OKADA and SUGINO [1937]. The numbered diagrams on the right of the figures are original.

Type I

Two fragments taken at the worm's two extremities are joined, which means that a simple gap is created in the anteroposterior system. For example, region *a. p* is joined to region *p. p*. This means that levels 6 and 11 are put side by side. Consequently, levels 7–10 are to be reestablished. The worm recovers a normal organization.

Type II

Two identical fragments are grafted by their homologous edges. For example, two fragments *a. p* are joined by their posterior amputation surfaces. They are brought together at level 6 whose cell equilibrium is identical. Neither can supply new extracellular information to the other. Consequently there will be no intercalary regeneration. Exceptionally, supernumerary end pieces (head or tail) can form laterally but only if the two fragments are imperfectly joined and regenerate independently of each other.

Type III

In the preceding example, the absence of intercalary regeneration can be attributed to the opposing polarities which is inevitable if identical edges of two identical fragments are joined. To discard this interpretation, two fragments taken at different levels are joined after one of them has been turned round and their anteroposterior axes have opposite polarity. For example, $p.\,p$ and $a.\,p$ are joined at their anterior edges (levels 11 and 3). Intercalary regeneration is possible. It restores levels 4–10 and, as proof, a pharynx appears, whose polarity conforms to $p.\,p$.

Type IV

Two fragments are cut from the same planarian and the more posterior of the two is put in front of the other. For example, $a.\,p$ is grafted by its anterior edge to posterior edge of p. Level 3 is put behind level 10. A supplementary pharynx will appear in the newly formed part with *inverted polarity*. This result, paradoxical at first glance, is predictable if the concept that the whole sequence of intermediary levels between those that are joined must be reestablished is correct. In the present case, levels 4–9 are reestablished, but since 10 was put in front of 3 their sequence is inverted. SUGINO [1941, 1953] confirmed this inversion of polarity each time a more posterior level was placed in front of a more anterior one in the system. When no pharynx is rebuilt, the result can be tested by the beat of the epidermal cilia.

These relatively simple mechanisms of intercalary regeneration allow us to explain the various conditions of morphogenesis that are set off by grafting a fragment of a planarian on a whole specimen. In such cases the fact that one part of the system is represented twice has no effect: the nature and the size of the region produced by regeneration will depend upon the level of the graft bed on the host and the level of the region of the graft by which it will be grafted. These two levels will mutually modify the tissue equilibrium around the wound. Intercalary regeneration can manifest itself as a lengthening of the graft or as an alteration of the receiver of the graft; sometimes the origin of the regenerated material cannot be determined. When heads [SCHEWTSCHENKO, 1937; TESHIROGI, 1956] or parts of the head [OKADA and SUGINO, 1937] are grafted in the prepharyngeal region which is very close to them in the normal integration system, they produce practically no changes. But, if they are grafted in the postpharyngeal region, the part of the system between the graft and the level of their insertion includes many levels. A supplementary prepharyngeal and pharyngeal region will develop through intercalary regeneration. For the same reasons, caudal end pieces [SCHEWTSCHENKO, 1937] or fragments taken from the tail [OKADA and SUGINO, 1937] acquire a prepharyngeal region and a pharynx only if they are implanted in the prepharyngeal region. Prepharyngeal fragments implanted behind the pharynx also provoke the appearance of a supplementary pharynx. Even if the graft's polarity is inverted with respect to the host, intercalary regeneration can still take place.

The understanding of intercalary regeneration is fundamental. It not only clarifies a certain number of phenomena that cannot be explained by other morphogenetic concepts but it also allows us to reconcile experimental results that seem contradictory. A difference of opinion between BRØNDSTED [1939] and ANSEVIN [1969] is a good illustration. As we saw earlier (p. 99), when BRØNDSTED grafted a head on the posterior cut surface of a short fragment he observed neither the reorganization of the fragment nor the inhibition of the cephalic regenerate which normally developed on the anterior section. ANSEVIN first grafted the head on a posterior cut surface and made only later a more anterior transversal cut to produce a short fragment. Under these conditions, she observed inhibition of the anterior regenerate and the formation of a pharynx with inverted polarity. She attributes this result to the fact that the anterior cut was made at a time when the graft was already established, whereas in BRØNDSTED's experiment this was not the case. Yet when the regenerated head in BRØNDSTED's experiment was amputated again, the situation became the same as in ANSEVIN's experiment, and yet a normal head was formed a second time on the anterior surface. This contradiction can be explained by intercalary regeneration. In fact, *the fragments used for the experiments do not belong to the same levels.* BRØNDSTED's material comes from the prepharyngeal region. Between this and the head grafted on the posterior surface, there is only a very small portion of the anteroposterior system. Consequently there is no appreciable intercalary regeneration. In particular, the pharyngeal region is not included among the levels that are to be reconstituted. ANSEVIN, on the other hand, took her fragments from the postpharyngeal region. Thus, in these cases the prepharyngeal and pharyngeal levels had to be reestablished between the postpharyngeal region and the grafted head. (Incidentally, SENGEL [1953] demonstrated earlier that grafting a head near the caudal end of a planarian can produce a pharynx with inverted polarity; he attributed this to 'induction'.)

We now have to return to more common regeneration phenomena, for example those found after a simple amputation, which, at first glance, may seem quite different from intercalary regeneration. Theoretically, if a level is deprived of neighboring cells (this happens when one amputates one of the worm's end pieces), the tissues are freed from the dominance which until then was imposed upon them; they cannot receive new information allowing them to replace the missing levels. The only transformation that offers a way out of this situation is a displacement of differentiated cells. Such a displacement introduces new patterns – those of a distant level in the normal system – and intercalary regeneration follows automatically. We shall see in the following sections that this effectively occurs. Displacement of epidermal cells during wound closure which reestablishes the distal level of the regenerate is a prerequisite of blastema formation. In accordance with the system of graphs defined above, when an amputation removed levels 1–5, level 1 is reestablished during wound closure and afterwards levels 2–5 follow in succession. Thus, the main point needed for understanding regeneration after an amputation is found in this phenomenon of intercalary regeneration which was well known for a long time but whose theoretical significance was underestimated.

We may conclude with formulating the laws of intercalary regeneration, as deduced from OKADA's und SUGINO's work:

First Law

When two morphogenetic levels, belonging to the same integration system but situated far apart, are brought next to each other the one modifies the tissue equilibrium of the other. This will start an intercalary regeneration which will stop when the sequence of levels *between the two levels existing in a normal system* is completed. This phenomenon is independent from portions of the system adjacent to the levels which are joined together. It takes place between regions no matter whether they have opposite polarity or identical polarity.

Second Law

When two pieces with opposite polarity are joined at identical levels, neither of the two receives new information that can modify the tissue equilibrium. Consequently, there is no intercalary regeneration. However, the two connected parts can rebuild missing end pieces if their surfaces are imperfectly joined.

Third Law

If a part of an anteroposterior system is placed in front of a more anterior part of the same system with identical polarity, the sequence of levels formed by intercalary regeneration is inverted. *Thus, the inversion of polarity in newly formed tissue is not due to an alteration of the polarity of cells or tissues in the orginal parts but to an epigenetic process in the regenerate.*

The Progression of Differentiation in the Regenerate

The development of the blastema is independent of the proliferation of dedifferentiated material. The levels are outlined according to a basipetal sequence, then morphogenesis progresses more or less in the same way as in the embryo; differentiation involves typical inductions and regulation allows a complete regenerate to form if the blastema is deficient. Both in anterior and in posterior regeneration, morphallaxis immediately follows epimorphosis and establishes more proximal levels.

One of the most striking aspects of planarian regeneration is the fact that the time span needed to rebuild the various organs, especially the eyes, is always the same in regenerates which are developing at the same level. Thus an amputation triggers a 'developmental clock' that is as precise as that of embryogenesis. But in planarian regeneration it seems to depend solely on differentiation and not on cell proliferation. At least this is what can be inferred from counting mitoses in fragments of the same size, coming from the same batch and regenerating at the same speed: in the syncytium the maximum number of mitoses usually exceeds the minimum number 3 or 4 times [CHANDEBOIS, 1972]. We have already seen earlier that planarians decapitated immediately after X-ray treatment can regenerate their head without any mitoses. Experiments carried out by GRÉGOIRE [CHANDEBOIS, 1968b] with *D. lugubris* are even more significant. If in this species the amputation is made sooner than 7 days after irradiation, regeneration is still possible, but the later the amputation, the smaller will be the regenerate obtained. In regenerating worms, 15–20 days after irradiation necroses appear in the ventral epidermis near the genital orifice. The planarian dies 20–30 days after irradiation. Mitotic activity has been measured in worms decapitated just after irradiation and found completely inhibited as early as the day following irradiation and amputation. But it may be restored and mitoses may be observed 10 days after irradiation, that is, just when the ability to form a blastema disappears. The mitotic activity increases during the following days, when the first signs of necrosis appear. Thus, mitosis and regeneration are quite independent and unrelated in this species.

The speed with which organs in the regenerate appear varies in the same individual with the level at which the amputation was performed. It is a well-known fact that the time needed for eye differentiation increases as the level of amputation is farther and farther away from the head [ABELOOS, 1930; SIVICKIS, 1931; DUBOIS, 1949; BRØNDSTED, 1942]. This time lag seems to be set during the programming of the regeneration process. In fact, when cephalic blastemas are isolated *in vitro* their missing eyes reappear later if the amputation was made at a more posterior level [C. SENGEL, 1963]. This gradation is not related to the quantity of undifferentiated material available. In fact, BERMOND [CHANDEBOIS, 1968b] counted mitoses in individuals of *P. cornuta* and *D. tigrina* cut into equal fragments: two prepharyngeal (B1 and B2), two pharyngeal (C1 and C2) and two postpharyngeal (D1 and D2). In each individual the maximum number of mitoses in the syncytium (as well as in the type I cell system) was found in fragment C1. The average number of mitoses in the syncytium in B1 and D2, respectively, was 46 and 33 % of that in C1. KRITCHINSKAYA and LENICQUE [1969] came to a similar conclusion: the rate of regeneration bears no relation to the number and the density of 'neoblasts'. Thus, the developmental clock is not set in the same way for blastemas formed at different levels of the body. As yet there is no explanation for this phenomenon. The author thinks that it is probably just a spatiotemporal pattern, having nothing to do with the determination of regenerative morphogenesis. On the contrary, the existence of a 'time-graded field' is considered by BRØNDSTED since 1955 as the clue to the problems of planarian regeneration.

Here some considerations concerning the workings of the developmental clock during epimorphosis and the subsequent morphallaxis are in order.

Epimorphosis

The blastema is usually considered to be a mass of undifferentiated cells which are waiting for a 'signal' from the stump to start their differentiation. According to WOLFF this signal is the establishment of a sort of organization center, the brain, that would be the starting point for epigenesis in the regenerate. This interpretation seems to be supported by current histology, because in the young blastema all the elements look alike. However, if this view were correct, one should be able to point out a stage, early in regeneration, when the blastema is indeed completely undifferentiated. But C. SENGEL [1960, 1963] has shown that as soon as blastemas are large enough to be isolated for culture *in vitro,* they already possess the capacity to form all the structures they would have formed if left undisturbed. Blastemas from *D. lugubris,* formed 3 days after amputation were explanted. The head blastema acquires a brain and then eyes, by 8–15 days in culture. The tail blastemas acquire tail structures. If a head and a tail blastema are joined orthotopically, there will be an intercalary regeneration and a pharynx forms. The conclusion must be that the young blastemas already represent an organized whole; they are determined and their polarity is set.

The outgrowth of a blastema does not mean that there is an accumulation of undifferentiated cells which will be later exposed to induction as a whole. The outgrowth of the blastema coincides with the start of the developmental clock in the first dedifferentiated elements produced in the stump near the site of amputation. Observations with electron microscope [CHANDEBOIS, 1973a], already referred to earlier, showed that the cells of a 3-day-old blastema, in which eyes have not yet been formed, are not undifferentiated. Their differentiation simply has not yet reached a degree to be noticeable by gross morphological observation or even with routine histological techniques. Moreover, the degree of differentiation is not the same in all the cells and *varies according to an axial gradient.* At the apex, the tissue more or less resembles that of the stump. Its very irregular shaped nuclei are of different types. As one gradually approaches the base of the regenerate, the cell types become more uniform. Especially the nuclei, ovoid and clear, look very much alike. The syncytium, with many free ribosomes, very few mitochondria and little reticulum, is observed only at the level of the amputation. There one notices cell territories that are individualizing. In the stump, near the level of amputation, there are dedifferentiation phenomena and one finds the only elements of syncytium that are in mitosis. This gradient in the process of differentiation indicates that the levels are set up in a *basipetal* sequence.

This direct proof, obtained by electron microscopy, confirms the conclusions reached by experimental work done on the marine planarian *P.*

lobata [CHANDEBOIS, 1950, 1951, 1957a]. The specimens amputated in the prepharyngeal region and regenerating in hypertonic solution (15 g NaCl/l seawater) very frequently produce asymmetrical heads, up to 100 % in certain series (fig. 42A). Some have only a single eye and one side is completely missing or is represented only by a kind of undifferentiated border that goes along the median coecum. Others have two eyes, located at exactly the same level, but one side lacks most of the distal parts. These asymmetrical heads form a continuous series in which the anterior edge of the incomplete half-head is situated at different levels of the anteroposterior axis. These heads rapidly become symmetrical when returned to normal seawater (fig. 42E). The anomalies produced are related to the fact that the transversal sections involve both the parenchyma and the diverticles of the coecum. The presence of gastroderm on the site of amputation represents a local obstacle to prompt healing. In normal seawater, it will delay the healing only slightly and its effects are counteracted by a regulation that takes place very early in the blastema. But the hypertonic solution increases this time lag and prevents regulation. The greater the delay is on one side of the amputation surface, the greater will be the deficiency of the retarded part of the regenerate (1–5 days for the heads with two eyes; 4–13 days for asymmetrical heads with one eye). If a new cut is made at a level more anterior than the base of the retarded half-head, the removed part is regenerated but the retarded half-head is not. In contrast, if the cut is made exactly at the level of the anterior edge of the retarded half-head, the latter will complete itself, but the non-defective side will not regenerate, so that the asymmetry is reversed (fig. 42D). These results show that the part of the cut surface which heals first produces a blastema whose differentiation starts with the setting up of the distal end and progresses regularly towards the base. When the blastema appears on the retarded cut surface, the other half has already differentiated a number of the morphogenetic levels of the anteroposterior system. These levels, because of an inhibitory action, will not be reproduced in the retarded blastema. The latter can only produce more posterior levels. This same series of anomalies is found in caudal regeneration and in regeneration along the sagittal incisions (fig. 42B, C). Thus, differentiation progresses in a basipetal direction in all types of blastema.

The fact that the levels are successively determined does not mean that the factors in charge of this determination will determine the role of each individual cell in reconstitution. A morphogenetic level is simply a primordium; its definitive organization will take place later, with the help

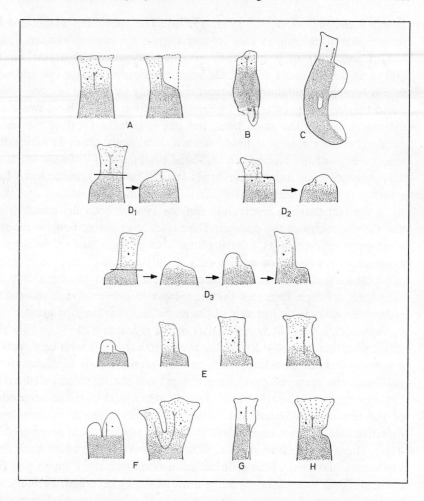

Fig. 42. Asymmetrical regeneration in *P. lobata*. *A* Asymmetrical heads with one and two eyes. *B* Asymmetrical tail. *C* Incomplete regeneration in the left half of a beheaded worm. *D* Regeneration of asymmetrical heads: the asymmetry is reversed when an incision is made exactly at the level of the distal end of the retarded part (2 and 3) but not when it is made more anteriorly (1). *E* Regulation of an asymmetrical head. *F* Regulation of two half-heads formed on the same section. *G* Regulation of a half-head in the left half of a decapitated worm which has not regenerated along the sagittal cut. *H* Head whose half formed by regulation is not blended with the stump. From CHANDEBOIS [1957].

of cell interactions, entirely comparable to embryonic inductions. LENDER [1952] clearly demonstrated the brain's role in the differentiation of the eyes in *P. nigra*. Cells become competent and are induced following a spatiotemporal pattern established during the regenerate's determination but independent of the order that established the levels. CORNEC and FONTAINE [1966], working in the author's laboratory, have shown that in *P. cornuta* not all the eyes regenerate simultaneously. First the eyes appear on the left side of the head, then 24 h later, the right side eyes follow. The anterior edge, although it is the first to be determined, will form eyes 1 day later. These differences come from the tissues' intrinsic properties. In the same batch of *Polycelis,* right or left half-heads were removed by two perpendicular incisions. The right half-heads regenerated eyes in the left corner, 24 h earlier than the left half-heads, in their right corner.

Since the young regenerate's cells are not yet fully determined, but are still involved in the epigenetic process resulting in their definitive differentiation, regulation is possible in cases of deficiency. In *P. lobata,* a necrotic area sometimes remains in the middle of the cut surface. Consequently, the blastema splits in two. Each half-blastema forms a half-head and each half-head reforms its missing part by lateral development (fig. 42F). If *Procerodes* specimens are decapitated, then split in two by a longitudinal incision, each half first forms a half head, which then completes itself, even if regeneration did not take place on the sagittal cut surface (fig. 42G). We observed in this species, that in some cases the half of an asymmetrical head which was formed by regulation cannot join up with the retarded half-section although it grows transversally. It stays separated from it by a notch (fig. 42H). In these cases, the restoration of the missing part is obviously carried out by the regenerate alone, independently of all intervention from the stump. In freshwater planarians, regulation is very precocious and represents a reorganization of the whole regenerating part. For example, head blastemas formed on a part of the cross section of the body or on a notch are symmetric from the beginning. However, the two eyes are not differentiated simultaneously; the eye situated on the same side on which the partial incision is made is ahead of the other by a few hours [ABELOOS, 1930]. Thus, the presumptive fate of the early blastema is to provide the side of the regenerate that corresponds to its base. But this fate is not definitely determined. It tends to produce a whole head. Late regulation takes place in an already organized regenerate. It is carried out like an ordinary regeneration needing 5–14 days. In contrast, precocious regulation in the apparently undifferentiated blastema is identical to regulation in young embryos; in case of a deficiency, some of the cells change their course of development so that at a later stage, the blastema turns out to be complete and normal.

The conclusion which can be drawn from these facts is that the cut surface functions as a generating layer. As soon as the dedifferentiation (often followed by multiplication) produces the first elements of the syncytium, these will immediately start to redifferentiate. They are pushed forward by elements produced later which come between them and the stump. These younger elements also differentiate somewhat later and form a more proximal level than the previous ones. Dedifferentiation continues in the stump so that more and more proximal levels are added to the blastema one after the other. This initial differentiation is a preparation to the later inductions needed to give definitive potencies to the primordium (certain cells become inductors and others enter a competent state to receive induction). This later phase makes regulation possible in case of deficiency.

Morphallaxis

Morphallaxis follows epimorphosis. It can be most easily recognized when there is a reconstitution of a pharynx, either behind head regenerates formed on the postpharyngeal cut surfaces or in front of tail regenerates formed on prepharyngeal surfaces. In terms of current concepts, *only* the dominant zone, that is the head, can order the reconstitution of intermediary parts. According to WOLFF [1962], for example, the 'inductions' can proceed only in succession from head to tail. Thus after the reconstitution of a head, the epimorphic regenerate elicits morphallaxis. In the same way, a caudal end piece reconstructed by epimorphosis remains passive and the morphallaxis that completes it is induced by the head. This leads to the view that differentiation does not take place in the same way in anterior and posterior regeneration. WOLFF et al. [1958] tried to support this view by destroying blastemas with X-rays. The parts that did not have a cephalic end (and could not reconstitute it) were the only ones that did not form a pharynx. But these results are less convincing when one realizes that the fragments, which had a cephalic end piece, also had the base of a pharynx that regenerated on its own, despite the destruction of the blastema. After destroying the tail blastema in cephalic fragments isolated by a prepharyngeal amputation, a pharynx never appears [CHANDEBOIS, 1959].

The marine planarian *P. lobata* provides a whole series of anomalies whose observation alone is enough to prove that morphallaxis is set off by epimorphosis, whether the regeneration is anterior or posterior [CHANDEBOIS, 1957b]. The best example is heteromorphosis. When a tail regenerate is formed on an anterior cut surface at the prepharyngeal level, a pharynx with inverted polarity will be formed through morphallaxis (fig. 5), exactly as if a caudal extremity had been grafted on the wound (cf. case No. III of intercalary regeneration, p. 111). Moreover, when the anomalies occur during the determination of a blastema, they are felt throughout the regenerated part. For example, when asymmetric heads form an amputation surface near the pharynx, morphallaxis affects the whole width of the body in the case of asymmetrical heads with two eyes. But behind asymmetric heads with only one eye, the stump stretches only in the corresponding half of the body. That half of the cut surface which did not produce a blastema will move progressively farther away from the anterior end. The same phenomenon is observed in posterior regeneration. When a tail blastema is asymmetrical, the pharyngeal zone is usually also asymmetrical. The median part of the regenerate may not differentiate, as a result of contractions along the wound's edges. The two coecums are abnormally close to each other. In these cases, the median part of the pharyngeal zone, including the pharynx, is not reconstructed [CHANDEBOIS, 1957].

Thus it seems that differentiation progresses exactly in the same way in anterior and posterior regenerates. It starts with the establishment of the most distal level, then the more proximal levels are reconstructed. The process first uses only dedifferentiated elements, then it turns to another histogenetic mechanism: the displacement of the equilibrium of cells in the stump's cell transformation system. One can thus consider the succession epimorphosis-morphallaxis to be a simple manifestation of the law of basipetal progress in the determination of regenerates.

The Role of Healing in Regeneration

The meeting of dorsal and ventral epidermis during healing is in itself a necessary and sufficient condition for the formation of a blastema, representing the anterior or posterior end of the worm. Before a head forms, the dorsal epidermis redifferentiates when it comes into contact with ventral epidermis; the opposite happens before the formation of a tail. The resulting restitution of a distal end sets off an intercalary regeneration.

Is the establishing of a distal level only the first stage of morphogenesis, or is it a determining factor for the beginning of regeneration? *In vitro* culture of fragments of *D. gonocephala* and *D. subtentaculata* [CHANDEBOIS, 1968a] have shown that dedifferentiation and proliferation *per se* do not automatically lead to the formation of a regenerate. Generally speaking, wound healing does not occur in fragments isolated in the prepharyngeal zone by two cuts and placed in a medium suitable for histiotypic culture (p. 80). In some cases, epidermis coils and then the fragment immediately begins to desintegrate. Usually gastrodermal cells swell and seal the wound. In these fragments dedifferentiation is observed. Mitotic counts have shown that divisions of syncytium nuclei are more numerous than in regenerating controls. There they proved that the medium keeps the parenchyma in good physiological condition, but there is no attempt in the fragment to restore the missing structures. Histological controls suggest that the syncytium does not differentiate. Sooner or later, in some cultures, fragments become capable of healing. The anterior wound is more frequently closed than the posterior one because the gastrodermis is less protruded and muscle contractions make it shrink. Healing occurs on the posterior wound where gastrodermis is eliminated. In every case normal blastemas are differentiated even when closure occurs only more than 10 days after explantation.

Observations have shown that not all the closing processes which restore epidermal continuity can be the starting point for regeneration. On amputation surfaces which cut through the base of the pharynx there is often no sign of a blastema [CHANDEBOIS, 1957a]. Yet undifferentiated material is not lacking since mitotic activity in the syncytium is higher than at other levels [BERMOND, in CHANDEBOIS, 1968b]. Each time regeneration failed to occur, the cut surface noticeably contracted right after the amputation, became V-shaped and then closed. The dorsal epidermis of one half of the wound fused with the dorsal epidermis of the other half; the ventral epidermis reacted in the same way. Observations of explants cultivated *in vitro* point in the same direction. Fragments were taken from a morphallaxis zone where there was a rich syncytium [CHANDEBOIS, 1968a]. When the dorsal epidermis was injured during explantation, more or less spherical, whitish masses appeared through the openings. Histological study has shown that in these masses the epidermis is poorly differentiated. Despite the reestablishment of epidermal continuity, there were no traces of organization. The masses that appeared consisted of a more or less syncytial parenchyma, and looked more like tumors than blastemas.

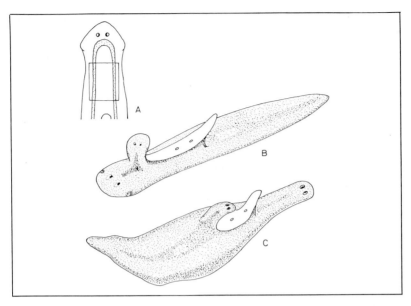

Fig. 43. Experiment of SCHILT [1970]. *A* Diagram of the operation procedure: the fragment is reimplanted with its dorsoventral polarity reversed. *B* The additional little planarian formed when the fragment is reimplanted with its anteroposterior polarity conforming to the stump's. *C* The additional little planarian formed when the fragment is reimplanted with its anteroposterior polarity reversed.

Each time regeneration takes place, the dorsal and ventral epidermis are confronted on the transection's surface. This way of closing the wound is truly a determining factor in the formation of a blastema, for the experimental joining dorsal and ventral epidermis, without tissue loss, is enough to provoke the appearance of supplementary heads or tails. The experiments of STÉPHAN and SCHILT [1966, 1967] complemented by SCHILT [1968, 1970] illustrate this very well although these experiments were not interpreted by their authors (fig. 43). A rectangular fragment was cut out of the prepharyngeal or pharyngeal region and reimplanted *in situ*. In one set of experiments, they were in normal orientation but in the other series the anteroposterior axis was rotated by 90 or 180°. The rotation had no effect; both the rotated and the nonrotated grafts healed and the worm remained normal. In another set of experiments, the graft's dorsoventral polarity was inverted, while the anteroposterior axis was or was not modified. In these experiments the graft's dorsal epidermis joins up with the host's ventral epidermis and vice versa. In such experiments at the graft's anterior edge (anterior before rotation) a head emerges and on the opposite edge a tail forms. Lateral edges develop into lateral borders with adhesive bands. Thus a complete little planarian emerges from this regeneration. Its back is fused with the host's back and its anteroposterior axis either coincides with or is perpendicular to the axis of the host. Sometimes another small planarian is formed on the ventral side of the host. STÉPHAN and SCHILT excluded the possibility of the intervention of the nervous system in the determination of these supernumerary formations, for the results are the same when the grafts are taken from the area between the two ventral nerve cords.

The decisive role of wound healing can be explained by the cell transformation system. What the dedifferentiated cells actually do in the process of redifferentiation is necessarily controlled by *the extracellular information they receive at that moment*. In order to be able to produce new cell types and especially organize new structures, the *position of a certain number of differentiated cells must be changed* between the time of dedifferentiation and the process of redifferentiation. One easily understands that without these displacements, the only modification that the cells near the wound can undergo is to continue dedifferentiating and redifferentiating to keep up the existing structures. The healing that is indispensable for blastema formation necessarily involves the migration of epidermal cells that do not completely dedifferentiate. They obviously bring new extracellular information to the elements of the syncytium, which were up till then in contact with the exterior. With this in mind, one can understand that healing brings about cell equilibrium characteristic of a distal level since the end pieces of the body (head and tail) are essentially, from a topographical point of view, the meeting points of dorsal and ventral epidermis.

If indeed the distal part of the regenerate is established during healing, this certainly implies that the closing of a wound in the case of head formation is different from the same process leading to the formation of a tail. New research had to be undertaken to verify this important point. Recently this was done by HIRN [unpubl. data] in the author's laboratory, using the marine planarian *C. hastata*. Samples were fixed with osmic acid at 2 % in diluted seawater, enclosed in araldite. Sections of medium thickness were stained with toluidine blue and observed by light microscopy. This is the only method by which the wound epidermis could be kept intact in all the samples. Observations were made of individuals whose head or tail was amputated at levels where absence of regeneration is very rare and they were fixed 18–36 h after amputation. In every case, either the dorsal or the ventral epidermis of the stump is bent towards the wound (fig. 44). It continues as a flat wound-covering epidermis, without basal membrane, cilia or rhabdites. Where this epidermis approaches the epidermis of the opposite side, the cells become higher and ciliated. There is no cellular figure of transition between these cells which are not yet completely redifferentiated and the adjacent differentiated epidermis of the stump. These observations indicate that only one epidermis, either the dorsal or the ventral, provides cells for the wound covering epidermis. These cells first undergo incomplete dedifferentiation then they progressively redifferentiate while they reach the stump's epidermis of the opposite side. In the protruded blastema they already form a typical epithelium. Sections made from samples fixed 36 h after amputation show that the blastema primordium protrudes under those cells which were the first to redifferentiate. In all decapitated animals, it is the dorsal epidermis which provides the wound-covering epithelium and the blastema emerges on the ventral edge. In animals whose tail was amputated, the ventral epidermis closes over the wound and the regenerate is shifted somewhat dorsally. Moreover, in C. hastata, one often obtains head heteromorphoses. If one cuts them off, leaving a thin band of regenerative tissue, a heteromorphic head will invariably grow again [HIRN, 1973]. Thus one can follow the healing of posterior transections that would grow heads. Healing takes place in the same manner as normally occurs in anterior amputations, i. e. wound epithelium coming from the dorsal epidermis. These facts show that the cephalic or caudal nature of a blastema can be recognized right from the start of healing. This agrees perfectly with the idea that the wound's closing brings about the conversion of dedifferentiated tissue into the structure of a head or a tail.

We still do not know whether or not other tissues participate together with the epidermis in the distalization of the cut surface.[7] Moreover, it is

7   Many authors have assumed that the nerve cord has a morphogenetic role but this was never convincingly demonstrated. Recently, SPERRY et al. [1973] and SPERRY and ANSEVIN [1975] grafting small pieces of planarians together in various situations, came to the conclusion that the main nerve cord inhibits head formation and 'induces (directly or indirectly) the differentiation of tissues of the different body regions and the formation of normal body proportions'. This is, however, improbable as a number of heads is formed on longitudinal cuts made along the nerve cord (fig. 4). It should be pointed out that the absence of regeneration may have many causes. Muscle contraction which makes distalization impossible seems to be the most frequent one and also the more underestimated.

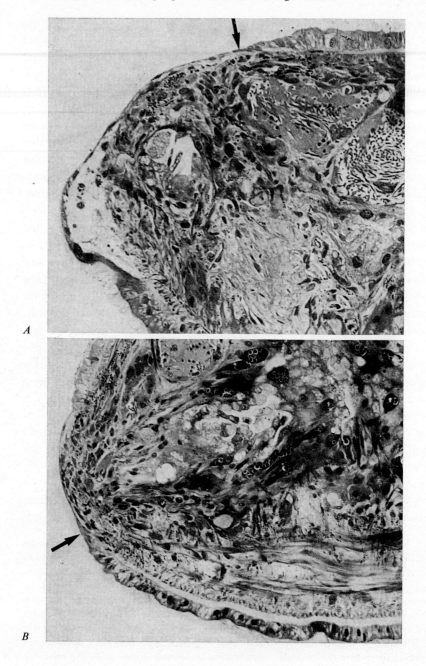

Fig. 44

known that in *G. abundans* [HAUSER, 1971] and perhaps in other species, the epidermis covering the cut surface is directly formed by the syncytium. It would be interesting to know if there are differences between anterior and posterior wound healing in this species.

The fact that a distal level is established creates, within the primary integration system, a situation comparable to the situation produced experimentally when a head is grafted on a pharyngeal amputation surface. This makes an intercalary regeneration inevitable. This regeneration stops when all the levels between the newly formed distal part and the level of the graft bed are reestablished, whatever the regenerate's polarity. In the same way, in the case of heteromorphosis the regenerate produces the normal sequence of all the levels between the amputation level and the distal end level which was reestablished first. This is demonstrated by the following facts. When the anterior part of the body is eliminated and the regeneration on the posterior part is heteromorphic (i. e. a tail end regenerates in place of the anterior end) a second pharynx will appear in the regenerate only if the amputation was made at a prepharyngeal level. No pharynx will regenerate when the amputation level is postpharyngeal. Similarly, when the posterior part of the body is eliminated and the anterior one regenerates in a heteromorphic fashion (i. e. a head end develops in place of the cut-off tail end), the regenerate will include a pharynx only if the amputation was made at a postpharyngeal level. There will be no pharynx in the regenerate if the amputation was made at a prepharyngeal level (fig. 5). Thus, in *P. lobata* heteromorphic tails include a pharynx only when they are produced by prepharyngeal sections but not by postpharyngeal ones [CHANDEBOIS, 1957b]. Heteromorphic heads, which are always formed on anterior stumps after prepharyngeal amputation are never followed by the fomation of an inverted pharynx. But in some freshwater planarians which reproduce by fission, a pharynx will appear in a heteromorphic posterior regenerate (head in place of tail) when short fragments are cut in the postpharyngeal zone. Consequently, the general rule is that a heteromorphosis is always a mirror image of the other regenerate (when the stump is a small fragment with two amputation surfaces).

Fig. 44. Healing before head and tail regeneration (1 day after amputation) in *C. hastata.* Arrows indicate epidermal cells spreading over the wound; dorsal cells in anterior regeneration (A), ventral cells in posterior regeneration (B). At the opposite side of the scar, the epidermal cells are thicker. Osmium fix. Araldite. Toluidine blue. ×90. From HIRN [unpubl.].

The Determination of Levels
in Intercalary Regeneration following Distalization

During intercalary regeneration which follows establishment of the distal ex-
tremity, each level is determined by the level formed before it. The stump remains
passive throughout this time: the regenerate either incorporates dedifferentiated ele-
ments (epimorphosis) or alters the cell equilibrium (morphallaxis) until all the miss-
ing levels are reconstituted. For this reason the stump cannot prevent the splitting-
up of structures that results from abnormal distalization (heteromorphosis, hemi-
heteromorphosis).

The final remaining question concerns the determination of levels during intercalary regeneration. Does the stump organize them, one by one? Or does the reconstruction of the distal level condition regeneration? In other words, is distalization the execution of the first part of the program of regeneration or is it the programming itself? *A priori,* only the first alternative seems to be acceptable since only the stump tissues contain differentiated cells, and are thus dominant over the dedifferentiated ones in the blastema. But if one accepts the concept of cell transformation systems, the second alternative is to be preferred, however, illogical this may seem at first glance. Let us suppose that an amputation was made at level 6; from it all regeneration takes place through epimorphosis. When healing has set up distal level 1, dedifferentiated material in the stump can only receive *new information* from this level (p. 109). By redifferentiating level 1 sets up level 2. Now level 2 supplies the new information to still available dedifferentiated material, which results in the formation of level 3, and so on. When the level 5 appears, it restores to level 6 its normal environment and regeneration stops. If the initial amputation is farther away from the distal end, making thus morphallaxis necessary[8], evidently the changes in the equilibrium of the stump's tissues cannot be determined by the stump itself, but only by neighboring newly formed levels. In this

8   Here the question could be raised: Why is morphallaxis necessary and why cannot regeneration be completed by epimorphosis? The answer is probably in mitotic events. In amphibians, morphallaxis is not necessary and does not happen because some dedifferentiated cells proliferate quickly near the apical ectodermal ridge. In planarians, as pointed out previously, mitoses occur in the stump and are not numerous. Only some cell types dedifferentiate (especially muscles). For producing a long regenerate, the animal would have to destroy a large amount of its cells. During morphallaxis, cell equilibrium is changed so that one level generates many (on a smaller scale) without returning to a completely dedifferentiated state. It may be that this is for the sake of economy rather than because of necessity.

view, the role of the stump is restricted to establishing a distal end level. Afterwards it passively accepts the influence of regenerated parts. More proximal levels will first incorporate its dedifferentiated elements and then attack the structures themselves, changing them according to the program of regeneration. To stop regeneration, the stump has nothing particular to do: the regenerate simply ceases to incorporate material when the last level formed finds itself in the same environment as in the normal integration system. If this theory is accepted, the regenerated parts should dominate the stump as soon as distalization is completed. This point of view could not be demonstrated if planarians always regenerated normally. Fortunately, the study of numerous anomalies, such as heteromorphosis and regeneration after asymmetrical healing, occurring in marine planarians, compensate for experiments that could not be performed.

Development of Regenerates with Inverted Polarity

Although heteromorphosis is a sporadic phenomenon, it does not take place completely by accident. If there is no experimental intervention other than a cut, it only happens on short fragments. It happens more often in short end pieces of the body than in pieces in-between. In such cases, the regenerated parts look exactly like the stump. In freshwater planarians this occurs only in the cephalic region. In some species, a head, isolated by a transection right behind the eyes, never forms a tail. In marine planarians, both the head and the tail end are more predisposed to producing heteromorphosis than short fragments taken from other levels. Thus, in *C. hastata* isolated cephalic end pieces produce more head heteromorphosis than short fragments of the prepharyngeal region. Heteromorphic tails are obtained only by accident from short fragments representing the posterior half of the postpharyngeal region. In *P. lobata,* there is a greater tendency to inverted polarity. Head heteromorphosis always occurs when isolated heads regenerate and happens accidentally in prepharyngeal fragments. Tail heteromorphosis occurs without exception on the anterior amputation surface of short fragments amputated behind the middle region of the pharynx and occasionally at prepharyngeal levels.

So far there is no sure technique to obtain heteromorphoses without fail in any species. One can only increase their incidence or make them appear, very infrequently, in types of fragments that do not produce them spontaneously, especially long fragments. The treatments used at the moment of amputation, or just before it, are so varied that they give no information about the mechanism of heteromorphosis. They are mitoclasic poisons [McWHINNIE, 1955; KANATANI, 1958], anesthetics like chloretone and ether [RUSTIA, 1924], partial inhibitors of metabolism like chloramphenicol [FLICKINGER, 1959], poisons of the nervous system like strychnine [CHANDEBOIS, 1957a], all sorts of other treatments whose effects on the cells is more or less complex and unknown: X-irradiation [HIRN, unpubl. data], LiCl treatment [TESHIROGI, 1955], blocking of regeneration for several days at low temperatures in a hypertonic solution or passing of an electric current [CHANDEBOIS, 1957a].

If one assumes that the stump determines each new level constituted in the regenerate, one must assume also that in the case of heteromorphosis the stump imposes inversion of polarity again and again on the cells as epimorphosis and morphallaxis progresses. If this would be so, it would mean that the polarity of the stump, or at least certain organs in the stump (for example, nerve fibers) are inverted. However, inversion is manifested only in the regenerated tissues. Since the body's end pieces can usually produce only one type of regenerate either head or tail, they appear as territories with 'partial morphogenetic potencies' [ABELOOS, 1932] comparable to the 'regeneration territories' of amphibians as defined by GUYÉNOT. Between the head and tail territories there is a bipotent zone where regeneration can go either way. The accidental production of polar heteromorphosis in this zone implies that the boundaries between these 'regeneration territories' are not precise.

The production of bipolar heads is still attributed to an alteration of metabolic gradients. Recently, RODRIGUEZ and FLICKINGER [1971] gave their assent to this interpretation. By treating fragments with chick embryonic extracts, they obtained bipolar heads, along with an increased synthesis of DNA. But head heteromorphosis cannot be attributed to this stimulation of DNA synthesis (i. e. mitosis) since mitostatic agents like X-rays or colchicine also act as stimulants of heteromorphosis [McWHINNIE, 1955; KANATANI, 1958]. Since COWARD et al. [1970] observed greater mitotic activity before the formation of a head blastema than before the formation of a tail blastema, it is possible that the modification observed by RODRIGUEZ and FLICKINGER is the result rather than the cause of the inversion of polarity.

After clarifying the way in which differentiation starts and progresses in a regenerate, the author put forward the hypothesis that the production of heteromorphosis is simply related to a *mistake in distalization* [CHANDEBOIS, 1973b]. Cells near the amputation (why, we still do not know), do not produce the topography of the end piece which they should reconstitute, but instead the topography of the other end piece. This does not necessarily involve permanent anomalies in the stump tissue, but can be explained just as well by accidental disturbances, localized in the vicinity of the amputation and taking place concomitant with healing. If the apical region determines the sequence of the new levels, one would understand that the stump cannot correct the mistake, and may even tolerate to be annexed to the heteromorphic regenerate by morphallaxis. The tissues of the end pieces of the worm (perhaps only their epidermis) may be more mistake prone than the tissues of the other levels so that they could produce only one type of distalization either head or tail.

To settle the differences between the two theories, HIRN [unpubl. data] worked out, in the author's laboratory, a preliminary experimental approach to the causal analysis of heteromorphoses. The research was done on the marine planarian *C. hastata*. Four types of fragments were taken from the anterior half of the worm: fragment I represented the cephalic end piece isolated by a cut at the level of the median cecum's anterior diverticle, and fragments II, III and IV each represented one third of the remaining prepharyngeal region. Under normal conditions, fragments I, II and III produce heteromorphosis fairly constantly (about 19, 8 and 7 %, respectively). Fragment IV, as well as postpharyngeal fragments, never produces any heteromorphosis. In this species, as in freshwater planarians, a more or less complete regeneration is possible after X-ray treatment, even if the dose is lethal (between 780 and 8,000 r), provided that the amputation is made within a certain time span after irradiation, which becomes shorter as the dose increases. After irradiation with 480 r, regeneration is blocked only if the amputation is made 14 days after irradiation. If one amputates immediately after irradiation the regenerate first develops, then on the 13th day the process stops and is resumed again after a pause, the length of which varies from case to case. Between 480 and 3,900 r, the regenerates develop slowly but sufficiently so that their nature (head or tail) can be identified without a doubt. In type I, II and III fragments from *Cercyra,* irradiated under these conditions and then immediately amputated, the frequency of head heteromorphosis is noticeably higher than in nonirradiated controls, but does not increase parallel with the dose. Thus, the frequency goes from 19 to 24 % in fragment I, 9 to 15 % in fragment II and 7 to 14 % in fragment III. Fragment IV, which never produces heteromorphosis in normal conditions, produces after irradiation about 7 % heteromorphoses. The same is true for longer fragments. Those representing I+II give 10 % heteromorphic heads; I+

II+III, 7.5 % and I+II+III+IV, 1 %. Evidently X-rays cause the posterior regenerate to develop into a head in a certain number of fragments that would have regenerated normally without irradiation.

After these results were obtained, further experiments were undertaken to determine at what time the irradiation was effective. Analogous experiments with varying the length of the time between the fragment's isolation and exposure to X-rays were made. If the X-ray treatment is applied 24 h before amputation, the percentage of heteromorphoses is the same as in animals where the irradiation was given immediately before amputation. But if irradiation takes place 48 h to 5 days earlier than amputation, the results are the same as for nonirradiated controls, that is, normal frequency for fragments I, II and III and no heteromorphosis for fragment IV and longer fragments. If irradiation takes place 24 h after amputation, it has no effect either. At this stage, as we have seen before, the wound healing epidermis has just been completed, but the blastema is not yet individualized. It will appear a few days later. This experiment shows that X-ray treatment can increase the proportion of heteromorphosis only if it takes place just before the healing starts. The time gap between irradiation and subsequent amputation must *not* be more than 24 h. If the irradiation takes place after the healing process is completed, there is no increase in the frequency of heteromorphosis. This proves that the regenerate is already determined. If irradiation takes place long before amputation (i. e. irradiation more than 24 h before amputation) its effect on the tissues disappears before the healing process starts. Thus, *healing itself is not affected by irradiation, but rather the stump's epidermis before it covers the wound.* It is assumed, as a working hypothesis, that if an amputation is made immediately after irradiation, the dorsal epidermis spreads over the wound's surface, while normally the ventral epidermis should do this. Inverted polarity will be the result. Unfortunately, this hypothesis cannot be verified experimentally because heteromorphosis does not always occur at any level. But there are some observations which indicate that the determination of the regenerate's polarity takes place in the early stages of the regeneration process, immediately after amputation, i. e. at the time of wound healing. When the stump – most probably its epidermis – recovers its normal condition, it can bring about normal regeneration again. However, the recovered tissues cannot correct any initial mistake which shows that the stump participates only in the establishment of the distal end of the regenerate and its polarity. All later regeneration processes are governed by the regenerate itself.

Other experiments carried out by HIRN [1973] confirm and complement these results. Heteromorphic regenerates produced by 60 fragments
I, II and III were removed again right at the base. Out of 20 fragments I
operated in this way, 13 regenerated, each with a heteromorphic head.
The fragments seem to be predisposed to inverted polarity. In contrast,
out of 20 fragments II, only one produced a heteromorphic head, 15
others regenerated a tail. Among 20 fragments III, the 16 survivors also
regenerated a tail each. Thus, the stump in prepharyngeal fragments –
with one exception – did not repeat its mistake during the second regeneration. This indicates once again the temporary nature of changes in tissue
properties determining heteromorphosis. However, two objections can be
made to this experiment. The first, that fragments cut a second time do
not regenerate under the same conditions as at the first time: the head
regenerate formed on the anterior section could 'dominate' the posterior
regenerate and determine it as a tail. To counter this objection, both head
regenerates were removed simultaneously. Of the 40 individuals operated
in this manner, none produced a heteromorphic head and 31 produced a
normal tail. The second objection that could be made is that the first regeneration could have renewed the stump's tissues and reestablished normal polarity in the stump in some mysterious way. Other experiments
done by HIRN [1973] show that this is not likely. Worms were cut into
left and right halves by sagittal section and were put in normal conditions
to regenerate. Thus, pairs of worms were obtained with identical heredity
and with very similar, if nonidentical, physiological states. Each individual
in each couple was cut into several fragments, as was done previously.
Despite previous regeneration, heteromorphosis occurred in proportions
barely lower than in the controls cut into pieces without previous regeneration, about 16 % for I, 6 % for II, 5 % for III. By studying the
distribution of heteromorphosis in these pairs of worms, HIRN showed that
reversal of polarity occurs completely at random and that its production
cannot be attributed to clearly defined morphogenetic territories in the
stump. It is quite exceptional that two homologous fragments of one pair
both produce a heteromorphosis (1 in 200 in I, 1 in 200 in II, none in
200 in III). The occurrence of heteromorphoses in fragments of the *same*
individual is quite without pattern. Sometimes only fragment I, sometimes
both I and II and sometimes even all three fragments regenerate heteromorphic heads. These cases could be explained by assuming some graded
variations of one single morphogenetic field for head formation in the
individual from which these fragments derived. But there are also cases

when fragment II or fragment III (or even both) regenerated heteromorphic heads but fragment I did not; or even cases where fragments I and III had heteromorphic heads while fragment II in-between them regenerated normally. No amount of speculation about a head-forming morphogenetic field in the stump can explain these cases. In the two individuals of a pair, the distribution of heteromorphoses among the various fragments is never the same. For example a worm that produced three heteromorphoses in I, II and III had a partner that did not produce any.

Though all these experiments shed little light on the nature of tissue modifications determining heteromorphosis, they do show that this anomaly is not the result of permanent and localized properties of well-defined morphogenetic fields. In fact, according to our theory, it is the consequence of a mistake in distalization, caused by abnormal healing. Since the changes in the stump tissues that provoke heteromorphosis (and are enhanced by X-rays) are transient and have no further effect once the wound healing is completed, it has to be assumed that differentiation is controlled by the regenerate's apical region and not the stump itself.

Heteromorphoses can now be explained relatively simply. When the stump makes a mistake in distalization, it establishes the most posterior level on an anterior amputation surface or the most anterior level on a posterior amputation surface. According to the third law of intercalary regeneration (p. 114), the consequence will be that polarity of the regenerate is reversed, without any change in the polarity of the cells or tissues of the stump. Thus, the fact that a number of epidermal cells are temporarily incapable of carrying out distalization in the correct fashion is enough to produce regeneration with inverted polarity from a normally polarized stump that is unable to correct the situation. This explanation is logically more satisfying than the earlier interpretations, since assuming that the origin of heteromorphosis stems from an inversion of the stump's polarity would make it necessary to ascribe mysterious properties to the tissues beyond the possibility of experimental proof.

As we mentioned earlier, heteromorphoses are sometimes accompanied by the formation of one or two supernumerary end pieces at the cut surface (fig. 5A, B). These are formed after the main regeneration is more or less completed and they are always tails in the case of head heteromorphosis and heads in the case of tail heteromorphosis. Their axis is perpendicular to the axis of the stump and is exactly at the level of amputation. However, according to the first law of intercalary regeneration (p. 113), when the regenerate's development reaches the morphogenetic level of the stump, morphogenesis should stop. One is tempted to compare the formation of supernumerary ends to cases described by OKADA and SUGINO [1937] where two identical parts grafted with inverted polarity, are not correctly joined and work together to form their missing parts. Complete duplication of the individual after heteromorphosis could indicate that the regenerate's transverse integration system is not correctly joined to the stump's system, which would mean that the former continues reorganization independently from the latter.

Asymmetrical Regenerations

When a regenerate is formed upon an incomplete base, it is deficient from the start. There are two types of such deficient asymmetrical regenerations. One type is due to a delayed wound healing, caused by obstruction of one part of the amputation surface by the outpouring of gastrodermal cells, either in a lateral (fig. 42A) or in a central position (fig. 42F). The other type occurs when the regenerate develops on a notch (fig. 3E). In such cases the regenerate formed on an incomplete amputation surface will be a partial head or a partial tail. But on the side of these partial structures is a missing complement that appears immediately after blastema formation and that grows rather quickly (except when regeneration occurs in hypertonic seawater in the case of *P. lobata*). Since this complement is formed independently from the stump, it is not always integrated with it through a secondary process. This is especially striking in the case of two heads that appear on the same section (fig. 42F) or when a regenerate develops on the edge of a notch (fig. 3E). In *P. lobata* [CHANDEBOIS, 1957a] the morphogenetic activity of the asymmetrical head's complement was analyzed in detail because of the variety of anomalies it produces when it fails to be integrated with the stump's transverse structural organization.

In a first type of anomalies which we called *tail hemiheteromorphosis,* an asymmetrical head and a completely symmetrical tail form on the anterior cut surface. Usually, the two regenerates form at about the same time. This is *primary hemiheteromorphosis* (fig. 45A), well known for a long time both in marine planarians [LUS, 1926] and in freshwater species after oblique amputations [MORGAN, 1904; BARDEEN, 1902; RAND and BOYDEN, 1913]. In this case the pharynx is formed by epimorphosis in the hemiheteromorphic tail regenerate. Between this tail and the stump, a supplementary head frequently develops, as happens after common tail heteromorphosis. In other cases, observed exclusively in *P. lobata* [CHANDEBOIS,

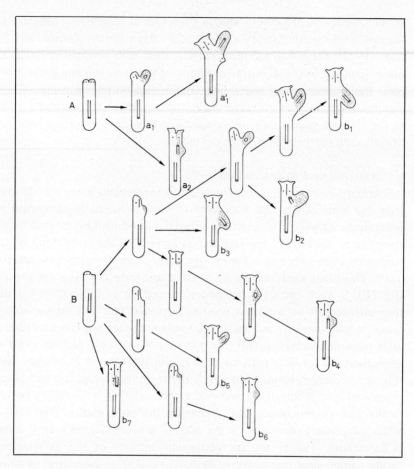

Fig. 45. Hemiheteromorphoses in *P. lobata. A* Primary hemiheteromorphoses. a₁, a′₁ = The two regenerates are kept separate; a₂ = they unite and a part of the tail forms the missing half of the head. *B* Secondary hemiheteromorphoses. b₁–b₃ = The tail is determined before head regulation; b₄–b₆ = the tail is abnormal because it develops when head regeneration is almost completed; b₇ = additional pharynx and eyes induced by the complementary head. Hemiheteromorphic tails are stippled. From CHANDEBOIS [1957].

1957a], the two regenerates do not appear simultaneously *(secondary hemiheteromorphosis).* The head *always* appears first and a supplementary head is never formed (fig. 45B). The tail can have normal or inverted polarity; the pharynx always forms by morphallaxis in the stump tissue. When the tail regenerate appears just before the end of regeneration of the head, it is very small and forms only a little outgrowth with no pharynx. Finally, the head regenerate in rare instances overlaps the tail regenerate during regulatory epimorphosis, so that behind the eye, formed on the side where wound healing and regeneration were delayed, a pharynx appears with inverted polarity and two ceca.

A hemiheteromorphic tail cannot be considered as a heteromorphic regenerate formed on a half-section. In fact, tail hemiheteromorphosis occurs in species – or at levels – where tail heteromorphosis never takes place. Moreover, the tail regenerate is symmetrical from the beginning, which shows that its formation is not initiated by the stump. There is no doubt as to the asymmetrical head's role in its determination. If the tail of a primary hemiheteromorphosis is cut, it regenerates, causing a secondary hemiheteromorphosis: the pharynx generally has normal polarity and is located behind the half-head that developed through regulatory epimorphosis (fig. 46A). Moreover, the head always develops first; the farther advanced it is in its regulation when regeneration starts on the retarded half surface, the smaller the tail regenerate and the greater the chance for having a pharynx with normal polarity (fig. 45B).

All these observations suggest that the complement of an asymmetrical head annexes at a relatively early stage in regeneration the material of the blastema formed on the slightly retarded half of the amputation surface so as to carry out a regeneration posteriorly (fig. 46C). The hemiheteromorphic half tail that is thus determined carries out in turn a regulation that completes it transversally. There is enough material for it to manifest its bilateral symmetry right from the start. But the external half of the tail is neither integrated with the stump nor with the head regenerate. Under favorable conditions, these two parts can manifest their regulatory power by causing a supernumerary head to develop belatedly. Thus, inversion of polarity in the tail regenerate is only superficial. In fact, the regulation phenomena proceed normally, but instead of repairing the regenerate's deficiency, the complement acts as the primordium of a supernumerary individual and the epimorphosis goes on until the individual is complete.

In other, even more conclusive cases, observed in *P. lobata,* the asymmetric head's complement reconstitutes the corresponding caudal part through morphallaxis (fig. 45B 7). The process is comparable to primary hemiheteromorphosis but it takes place inside the stump. The supernumerary pharynx has normal polarity and is situated on the retarded side. Supernumerary eyes can differentiate when ganglia are formed in

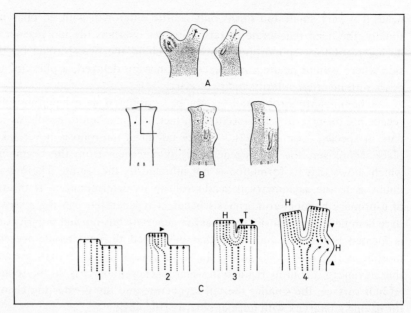

Fig. 46. Determinism of hemiheteromorphoses. *A* Secondary hemiheteromorphoses obtained after transection of the tail of a primary hemiheteromorphosis. *B* Formation of additional pharynges when worms regenerate in a strychnine solution. *C* Regulation in regenerating parts leading to primary hemiheteromorphosis. 1 = The left half of the cut heals and a half-head blastema is constructed under the influence of stump tissues; 2 = immediate regulation reestablished a complete transverse organization; 3 = the right half of the regenerated head is not integrated with the stump, its missing posterior parts are established in the blastema of the right half of the cut. This forms a tail in which organization is immediately symmetrical because again regulation occurs in transverse direction; 4 = the stump and the right half of the tail are both lacking anterior parts; they cooperate and form a second head; heavy arrows indicate direction of regulation. A, B from CHANDELOIS [1957]; C from CHANDEBOIS [1973b].

front of this pharynx. Once again, this clearly shows the tendency towards reduplication. The complement's role in the genesis of the pharynx was confirmed experimentally here, too. When one half of a head is removed by an L-shaped amputation, it is regenerated by the blastema that forms on the sagittal amputation surface. Its development resembles that of a complementary part in an asymmetrical regenerate. In specimens of *P. lobata* placed in a weak strychnine solution, such a regeneration is accompanied by the appearance of a pharynx in the stump on the operated side (fig. 46B). Asymmetrical regeneration usually does not produce such anomalies because the stump prevents it by a certain resistence to de-differentiation. This is probably due to the fact that the regions to be converted during morphallaxis are firmly integrated with the transversal structural system. One must therefore assume that the formation of a supernumerary pharynx will only occur when the tissues of the stump are somewhat weakened by the poison (or by starvation) and are more inclined to undergo dedifferentiation.

When the asymmetrical head fails to integrate the regenerate on the retarded side or establish integration with the stump, it will exert a certain inhibitory influence. During an asymmetrical regeneration in hypertonic solution, the levels of the anteroposterior system, already established on the advanced side when healing starts on the retarded side, will not be established on the retarded side. This is a clear case of inhibition. Its reversible character can be demonstrated by the removal of the advanced asymmetrical portion (including the complement). As we saw earlier (fig. 42D), after such an operation the retarded regenerate is completed antero-posteriorly and in turn prevents the regeneration of the amputated part. The head's complement alone is probably responsible for this inhibition since it includes regions that the retarded regenerate is unable to reform, both in the first asymmetrical regenerate and in the second one in which this complement is obviously present.

In summary, the complementary part of an asymmetrical head theoretically never finds in the stump's tissues near the level of amputation the cell equilibrium that could stop the automatic production of more posterior levels. This happens in the other half which is integrated with the stump. Furthermore, the complementary part can fuse with the retarded blastema and transform it into a tail. For this the stump produces all the material but it cannot prevent the duplication of the individual that takes place. When the opposite blastema resists assimilation, the complementary part may eventually expand at the expense of the tissues of the stump. The stump cannot block its conversion into anomalous structures and here again is a further source of dual individuality. Finally, if the retarded blastema develops into a head, the complementary part prevents it from forming an identical part. Thus, for all these anomalies caused by incomplete healing, only regenerated levels have the characteristics of dominance attributed earlier to the heads: inhibition of identical regions in the regenerated parts as a whole and 'induction' of supernumerary regions at the expense of regenerated parts or stump tissues. No comparable function could be found in the stump. Although stumps impose the asymmetry of regeneration at the start, they are powerless when faced with the consequences of the process of regeneration. At most, the stump can show a certain resistance to the tendency of the complementary part to assimilate stump tissues by morphallaxis.

Discussion

In worms, arthropods and amphibians regeneration presents itself as a distalization followed by an intercalary regeneration. The mechanisms of the progression of the regenerates' differentiation and their regulative power are reminiscent of properties encountered in comparable embryonic primordium.

On the basis of the data discussed in the preceding chapter, one can now propose a new concept for the mechanism of planarian regeneration. Although it is extremely simple, it applies to all types of normal (cephalic, caudal and lateral) and abnormal (heteromorphic, asymmetrical and multiple) regeneration. It may be formulated like this: the stump's only role is to give a morphogenetic program to the regenerate. The program is imprinted on the distal end of the regenerate, the first to be reestablished during wound closure. The imprinting is followed by an intercalary regeneration, guided by newly formed parts right to the end, while the stump provides only the cell material. In other words, the developmental clock starts to work and the contact with the stump stops the clock automatically when all the missing levels are reconstituted. A single action of the stump seems to be enough for programming all the regeneration. This action is carried out by the epidermal cells when they slide over the surface of the wound during healing. But this decisive action is also precarious. It can be impeded by other cells (muscle fibers that provoke contraction of the cut surface, gastrodermis that swells and prevents the spreading of healing epidermis) or it can take place in the wrong direction if the epidermal cells' migratory properties are abnormal (heteromorphosis). In these cases the developmental clock cannot start working or it starts with an incomplete or incorrect program. Of course, in the case of deficiency, the regulation initiated in the blastema usually reestablishes a normal situation, but like intercalary regeneration this is an automatic process which the stump cannot dominate. Furthermore, in cases where some regions already existing in the worm are repeated in the regenerated parts, they act like the primordium of a new individual that can develop right to a complete duplication of structures.

This concept is, as a whole, more in accordance with observed phenomena than concepts which include the idea of head dominance, because it covers a wider variety of observations and experimental results, some of which so far have found no explanation. To the author's knowledge, only one phenomenon remains still unexplained: the appearance of a pharynx in postpharyngeal fragments when there is no head regenerate. To obtain this result, ZILLER [1973] opened the healed wound every day. This is probably a rather special case because similar operations on cephalic pieces in *D. subtentaculata* [CHANDEBOIS, 1959] did not produce a comparable morphallaxis. Moreover, when no blastema is formed because of abnormal closing of the wound, the phenomenon does not seem to take place. One must, however, keep in mind that in ZILLER's experiments, the pharynx appears after repeated distalization, whose immediate effects are not yet known in detail.

Although the organization of the planarian body is rather different from other organisms, their dedifferentiated cells which, after an amputation, complete – or accidentally duplicate – the stump's structures, cannot be so fundamentally different from those of other animals. Therefore it is of significance that the concept that we have just outlined is consistent not only with the experimental results obtained with planarians, but also with the general patterns of regeneration in all group whose regeneration is thoroughly investigated (worms, amphibians, arthropods). After reviewing the literature (mostly the recent one) on these various groups, it seemed safe to conclude that the concept presented here could be used as a basis for a general theory of regeneration in metazoans [CHANDEBOIS, 1973b]. It is not feasible to go into details of the theory here but the main arguments justifying this generalization should be briefly stated, not only to corroborate the various conclusions of this chapter, but also to emphasize the importance of planarians in the understanding of adult morphogenesis.

(1) Traumatic regeneration (characterized by the establishment of a blastema that can be followed by morphallaxis) stems from particular properties of adult tissue in which cells help each other to maintain their differentiation, acquired during ontogenesis. Thus, regeneration appears fairly late in ontogenesis, when the definitive tissue dynamics are being set up. For this, the tissues have to reach a certain level of differentiation. Regeneration must not be mistaken for embryonic regulation, which is observed at stages when the primordia are not yet completely formed. Thus, in the case of the amphibian tail, regeneration is not possible during

embryonic development [NEWTH, 1958], while the planarians' head cannot regenerate before the differentiation of the nervous system [BARDEEN,
1902].

The fact that the tissue dynamics of the adult must be taken into
consideration to explain regeneration phenomena, illustrates the importance of the idea of a cell transformation system where all differentiated
elements influence, and are influenced by, neighboring elements. Yet the
idea of dominance exerted *from a distance* by one or more privileged regions is still accepted in all groups. It could only be contradicted in the
case of planarians because one could gather a greater amount of information. In these worms, regeneration takes place in small fragments in
anterior, posterior and lateral directions; the organization is sufficiently
complex for one to characterize fairly well circumscribed areas; the grafting is easy and allows all kinds of combinations among these areas; and
finally, the most important point, the regenerate can alter the stump. No
other group of metazoans offers so many possibilities for experimentation. In any case the idea of dominance from a distance and its supposed
mechanisms are still hypothetical. On the contrary, mutual aid among
differentiated cells is a well-known general fact. The resulting cell equilibrium is evident and perfectly explains all the aspects of regeneration. In
the great majority of metazoans there are no elements left that have retained histogenetic potencies. Thus one finds in the adult several cell
transformation systems, made up of reversibly differentiated cells (i. e.
mesoderm of the amphibian limb), or by tissue renewed by cells which
remain undifferentiated because they receive extracellular information
(vertebrate epidermis). Though the systems cannot exchange cell material,
the elements of each of them control the differentiation and the proliferation of the others. For example, one knows that in vertebrates, generally
speaking, the epidermis and its derivatives are necessary for maintaining
differentiation in the adjoining mesodermal elements. In its turn, the differentiated connective tissue restrains mitotic activity in the epidermis and
controls the differentiation of the epidermal derivatives. Thus relations between elements of two neighboring systems are set up in the same way as
relations within the same system. This explains why we could disregard
the segregation of histogenetic potencies in the account of the general
mechanisms of regeneration. But this simplification must not make one
forget the fundamental importance of the segregation of histogenetic potencies in a more detailed study of histogenesis and epigenesis.

It is interesting to note that BULLIÈRE [1971] proposed a scheme for the autoregulatory system of insect appendages, that is comparable to what CHANDEBOIS [1957a] proposed for planarians. Small regions are specified by morphological traits and the spatial disposition of their constituent cells. They are integrated in two spatial directions. If we reduce an appendage to the geometrical form of a cylinder and analyze it in mathematical terms, the morphological levels represent generatrices on the one hand and circles perpendicular to these on the other hand.

(2) Whenever, the study of a young regenerate was sufficiently detailed, the conclusion was always that the blastema – even in the earliest stages of its development – never represents a mass of undifferentiated cells. Thus, in amphibians, the limb buds, excised as soon as they appear, organize themselves autonomously when they are grafted onto neutral territory [KIORTSIS, 1953]. They already possess the necessary information for building up the limb's structures, which reminds us of the planarian regenerate's behavior *in vitro*.

It is known for all groups that the morphogenetic levels, along the axis of the regenerate, are not established all at the same time; this excludes *ipso facto* any global inductive action from the stump. In annelids the phenomenon can be easily observed because of metamerization. During cephalic regeneration, the prostomium is established first, then the peristomoium, and finally the more posterior segments. Levels that have just been reestablished do not immediately recover their definitive properties. They have to complete their differentiation and organize themselves, using cell interactions that have not yet been sufficiently explored, although some of them are known to be real inductions. Thus in a nemertean, SANDOZ [1965] showed induction of olfactory organs by the brain, which is the same process as the induction of eyes in planarians. The regenerates behave temporarily as a portion of a developing organism grafted onto the stump. The analogy with the amphibian limb is very striking [FABER, 1965, 1971]. The regeneration blastema has the same spatiotemporal patterns as the embryonic primordium. The differentiation of segments progresses in a basifugal direction: the most proximal one, the stylopodium, is already cartilaginous when the first finger is forming. As soon as it is formed, the young bud has an apical ectodermal ridge whose properties resemble those of an embryonic primordium. This apical ridge permits the limb to develop by stimulating mitoses in the adjoining mesoderm and organizing them into definite structures. THORNTON and THORNTON [1965] succeeded in removing the apical epidermis of the young blastema and grafted it to the base of the same blastema. The result was a duplicated regeneration;

the grafted apical tip and the tip that has regenerated in its place each imposed the reconstitution of the missing parts of the limb. Similarly, posterior regeneration in annelids can be summed up as the reconstitution of the prepygidial zone of growth that brought about metameres in the young animal.

The regenerates' regulatory properties that are due to their embryonic character and that are so remarkable in planarians are also found in amphibians. It is known that an incomplete base (for example, the stump of a limb split longitudinally) can produce a transversally complete autopodium [WEISS, 1926]. Regulation capacities are also manifested along the axis of the regenerate. DE BOTH [1970] amputated the limbs at the wrist. Several blastemas – which only had to regenerate the autopodium – were gathered and put in a cavity made under the skin of the flank. They joined to form a single primordium that produced a more or less complete limb. The stylopodium and the zygopodium were obviously determined without the stump's intervention, because of an embryonic regulation that was made possible by a sufficient quantity of blastemal material and the absence of dominance from the stump's tissues. In *E. tetraeadra,* a freshwater *Lumbricide,* COULOMB-GAY [1972] observed hemiheteromorphoses and showed that their determinism is the same as that of planarians. The anomaly stems from a deficient head regenerate formed on an incomplete base. When it undergoes a regulation, it produces a complete head which is not completely integrated with the stump and therefore continues the regulation as if it would be independent from the stump, producing eventually a tail.

(3) The study of planarians has shown that regeneration is set off by two successive disturbances in the cell equilibrium. The first is the result of the removal of the dominance of differentiated cells over cells near the transection. The second is caused by the displacement of differentiated cells during healing, which establishes near the level of amputation a topography analogous to that of the removed distal level in the tissue.

The stump's inability to reconstitute, even partially, the missing levels when healing does not take place is a well-known fact in amphibians. If the healing epidermis is removed each time as soon as it forms, dedifferentiation progressively reaches the stump's base. In this group, the reestablishment of epidermic continuity is not a decisive factor in determinig the formation of the regenerate. If one amputates a tail and then immediately puts epidermis on the wound, regeneration does not take place. But if the epidermis is pierced, there appear as many little tails as there are holes.

GOSS [1956] prevented the epidermis from spreading over the surface of leg stumps by implanting them into the body cavity. They underwent dedifferentiation but did not succeed in forming a regenerate. At most, a little cartilage formed adjacent to the stump's skeleton. During the normal regeneration of a limb, healing brings about distalization. This does not mean that the autopodium is the first to be reconstituted. But the healing epidermis differentiates the apical ectodermal ridge, analogous to the distal part of an embryonic primordium. In *Helleria brevicornis* (Crustacea Isopoda) HOARAU [1969] studied the regeneration of the lamellar expansions on the two sides of the pleon, the so-called 'neopleurons'. Even before the regenerate's development has started, shortly after healing, the bristle fringe that borders a normal neopleuron reappears. Thus the distal margin is the first to differentiate. In arthropods, distalization seems to take place as in planarians, by the meeting of heterologous epidermal regions. This is enough to set off the formation of a regenerate, even without removing any limb parts. By amputating insect appendages and reimplanting them *in situ* after rotating them, one can produce supernumerary appendages [BART, 1969]. Similarly, when openings are made in the neopleuron of *Helleria*, healing leads to the confrontation of dorsal and ventral hypodermis around the entire wound and, as a result, supernumerary neopleurons will grow [HOARAU, 1969].

Little is known about the role of wound healing in annelids. But COULOMB-GAY [unpubl. data] observed in *E. tetraedra* that the absence of regeneration is related to an abnormal wound closure process. Usually contractions shrink the wound, whose dorsal and ventral edges rejoin. In other cases, the contractions reduce the wound to a vertical slit and each level of the epidermis on one side fuses with the corresponding level on the other side. In these conditions no blastema is formed. In the posterior regeneration of annelids, the first metameres of the trunk to appear are the most proximal ones. However, this is preceded by a distalization that restores the pygidium.

One must note, though, that in amphibians and in annelids the joining of the two heterologous epidermal regions is not the only factor that can initiate regeneration. There are other circumstances that produce cell equilibrium analogous to that of a distal extremity. For example, it is known that the deviation of the limb's nerve in urodeles results in the formation of a supernumerary limb. However, the presence of a nerve is not the only determining factor in regeneration from a stump since YNTEMA [1959] obtained aneurogenic regeneration in limbs which were nerveless from the beginning. Similarly in annelids, the nerve cord was deviated to the body's lateral wall. This operation determines both the outgrowth and the nature of supernumerary end structures. An anterior part determines the formation of a tail, a posterior part that of a head. Here again the nerve cord is not the only factor that initiates the regeneration. AVEL [1932] obtained heads and tails without a nervous system. On the other hand, in *Syllis,* BOILLY and BOILLY-MARER [1972] showed that a deviation of the intestine suffices to produce a supernumerary tail, and yet a deficient tail can regenerate without endoderm [ABELOOS, 1950]. All these observations suggest that distalization can be achieved by different means in the same species and in the same type of regenerate. Is the same true for planarians? This is not impossible, but it would be difficult to prove since the organs in these worms cannot be dissected. The experiments of SCHILT (p. 123) show that healing may be sufficient. However, the anarchic production of heads (fig. 4) obtained when an incision is made along a nerve cord suggests that the nervous system may be also a determining factor.

(4) To determine the order in which levels are reconstituted in a regenerate, one usually had to observe very young blastemas or carry out experiments of the type performed in planarians. The results are not the same for all the groups; in some cases differentiation progresses basipetally, in others basifugally.

In arthropods, the sequence of determination is first observed on the basis of the size of the segment of the regenerate when the appendage emerges after moulting. The first formed are those whose size, at this stage, is closest to the definitive size. Thus determined regeneration seems to be basifugal for certain appendages, basipetal for others or even first basifugal then basipetal. However, NEEDHAM [1965] showed that a conclusion could not be drawn from such observations since the allometric growth coefficient of the various segments during regeneration could be different and not related to the sequence of determination. This point of view was recently confirmed by HOARAU [1973] in *H. brevicornis.* Histological sections were made of young regenerates of appendages that were not yet freed by moulting. While dedifferentiation goes on and reaches the stump's base, the regenerate acquires its primary shape by a remodeling of undifferentiated material. The segment borders appear very soon, which allows one to ascertain that the most proximal ones are the last to be formed. If one disregards the absence of mitoses in the stump and their presence in the regenerate, one has the outline of a mechanism that is identical with the one we proposed for planarians.

In amphibians, on the contrary, once distalization has occurred, differentiation progresses basifugally as in the embryo. Experiments confirmed the histological observations. In fact blastemas growing without an apical ectodermal ridge only organize proximal parts [STOCUM and DEARLOVE, 1972]. The later the operation, the more complete is the skeletal development. Far from presenting a contradiction, basifugal differentiation fits in well with the outline of the regeneration mechanism proposed here for planarians, this also holds for arthropods and for anterior regeneration of annelids. In fact, one has to consider the way in which undifferentiated cells are produced. Usually, as in planarians, they arise by multiplication of cells activated in the stump near the site of amputation. Elements which are just produced insert themselves between the stump and the elements which have already started to differentiate. Thus the most proximal levels are produced last. On the contrary, in amphibians the healing reestablishes an apical proliferation center and in annelids, along with the pygidium's differentiation, a growth zone is established. In these two cases, the newly differentiating cells insert themselves between the regenerate's distal end and the cells that differentiated before they did. Thus the first levels determined are the most proximal.

(5) Inversion of polarity is the anomaly most useful to understand the determinisms of normal regeneration. The comparison of results obtained in the different groups speaks strongly in favor of the concept here based on the study of planarians. In this group, intercalary regenerations make it possible to explain the inversion of polarity very simply. Polarity is inverted every time when a part of the body is grafted in front of a more anterior one (e. g. a postpharyngeal fragment in front of a prepharyngeal fragment). There is an inversion in the order of levels reconstituted, as shown by the direction of the pharynx and by the inversion of the ciliary beat in the epidermis. Similar phenomena were found in arthropods by recombining the segments of appendages in various ways. After intercalary regeneration, the polarity of the newly formed parts can be determined by the inclination of the bristles and spines. BULLIÈRE [1971], operating on *Blabera craniifer,* observed that when the sequence of levels of an appendage is modified so that some proximal ones are placed in a more distal situation (or vice versa) polarity is inverted in the intercalary regenerate. For example, this happens when the stump of an appendage amputated at the *tip* of the femur is joined to another appendage cut at the *base* of the femur. Thus, inverted polarity can be obtained *at will*; obviously it cannot be produced by just any change of tissue. It should be noted that the

sequence of reconstructed levels most likely determines the polarity of cells when they redifferentiate; but the polarity of cells does not determine the polarity of structures, as is generally considered.

The application of this law of intercalary regeneration to heteromorphosis is legitimized by the phenomenon on *inverse regeneration*. In amphibians, this was obtained by MONROY [1942]. A limb fragment containing the stylopode and the zygopode is grafted on the flank of the body sideways, so that the two amputation surfaces are free. The proximal wound heals; a regenerate with inverted polarity develops and it includes an autopode and a zygopode. In arthropodes, a slightly different operating procedure was used by BOHN [1965]. He cut out pieces of limbs and grafted them back *in situ* but with inverted polarity. Although the free amputation surface was proximal, it produced a distal part. Thus, whatever the orientation of the amputation surface, the epidermis during wound healing can only reestablish the appendage's distal end. When the amputation surface is more proximal with respect to the stump, distalization brings about the topographical conditions that inevitably lead to inverted polarity in the parts produced by intercalary regeneration. It all indicates that the inversion of polarity is not the result of an alteration in the stump's tissues, but the result of an abnormal sequence of morphological levels. The restriction of morphogenetic potencies characterizing the regeneration territory can be explained simply by the fact the epidermis can only reconstitute *one* kind of distal end structure.

The general outline of the mechanism of regeneration suggested by experiments on planarians can thus be applied to all metazoans. By carrying out *only one action* (the distalization of the amputation level during healing) the stump can program all the intercalary regeneration that follows. The missing levels are determined successively and their production stops automatically when the sequence is complete. Next, cellular interactions assure the progression of differentiation and the appearance of the organs. It is certainly somewhat disapointing not to be able to define more precisely the mechanism of the intercalary regeneration, which, from the observer's point of view, is the essential part of the repair capacity of the organism. Yet a more profound understanding of epigenetic phenomena would not be enough to satisfy our curiosity. The fundamental question is to find how the cells, belonging to hardly determined levels, use their genetic information to set up the species-specific patterns, and how, if their number is lower, they can often still perform their duty with the same accuracy. Moreover, this question can also be asked of embry-

onic development, for which one already knows the role of cellular interactions in the establishment of metabolic patterns characteristic for each
differentiated type. The determination of several levels in the blastula is
enough for these interactions to lead automatically to the production of
the specific morphogenetic patterns, even when some cell material is eliminated. If one sticks to facts and refuses all speculation, at present there is
no possible explanation. Thus intercalary regeneration, which one would
like to understand better is only one aspect of the tantalizing enigma that
molecular biologists handed over to morphogeneticists by teaching them
the translation of the genetic code.

Preliminary Report on a Metabolic Disease in Planarians

A lethal disease (with some cases of spontaneous recovery) can be inflicted on planarians by setting a wound without tissue loss. This disease does not occur after previous starvation and extensive regeneration. The disease strikes overfed worms spontaneously. Regenerative capacity disappears in sick worms like in irradiated ones. The disease is apparently related to cancer and seems to be provoked by a momentary overactivity of type I cells exceeding the animal's actual need.

It is difficult to believe that in an organism with such phenomenal regenerative capacities as a planarian a simple wound can have fatal consequences. Yet this is what was found in the author's laboratory. It is even more difficult to explain why this phenomenon remained unnoticed by investigators of planarian regeneration for such a long time. The details of the disease are still under investigation, but it will be useful and not out of place to present the preliminary findings in this monograph. The origin and development of the disease would be completely enigmatic on the basis of the older concepts of histogenesis and morphogenesis in planarians. But the new concept proposed here offers a reasonable explanation of the phenomena. At the same time, the study of the disease can also point out our lack of knowledge about metabolism and tissue dynamics of the planarians.

The disease was discovered by chance in the author's laboratory by mademoiselle MIREILLE MARTELLI in 1970. She tried to provoke tumors by inserting granules of scarlet red in the parenchyma, a procedure already used by SEILERN-ASPANG and KRATOCHWIL [1965] in freshwater planarians. When the powder proved to be ineffective, MARTELLI dissolved it in olive-oil and injected it with a micropipette. Instead of provoking tumors, the injections produced, after a time lapse, fatal lesions in all the species used for the experiment *(D. lugubris, P. cornuta, P. nigra)*. At first the worms looked normal, then after a certain time (varying from 4 to 22 days) the head started to regress. Death followed quickly in all cases. Since controls that had received pure olive-oil reacted similarly, various other liquids were used: liquid paraffin, a medium that had been used for culture of *Escherichia coli* and an azure blue solution. Regressions were invariably produced in all cases. At this point, MARTELLI tried injections with spring water, and then simply applied pinpricks. They also brought about the planarians' death in the same way. The experiments were continued by the author after Mlle MARTELLI's departure [CHANDEBOIS, in press]. Since it was clear that several simple manipulations can initiate the fatal consequences when water penetrates the tissues, it was thought to be simplest to set wounds with a knife without tissue loss and see if this also has the same effect. Although this seemed extremely unlikely it proved to be so. Thus a new channel was opened for research, all the more tempting since the phenomenon seemed at first inexplicable.

After a wound is set, the disease will manifest itself after an undetermined length of time – rarely over a month – during which the animal remains perfectly normal. One of the most characteristic symptoms of the disease is slow reduction of certain parts of the body which can lead to a more or less complete disappearance of these parts. The head disappears most frequently. The epidermis shrinks, usually without breaking (fig. 49II, A). The eyes and pigment are retracted to the prepharyngeal region. The worm gives the impression of having had its head amputated and the wound healed (fig. 47F, H). Later on, the scar tissue yields to inner pressure from necrotized tissues which produce a generalized swelling. If the worm does not die immediately, it is reduced to a postpharyngeal fragment incapable of regenerating. The tail can also be resorbed in this way. The pharynx also becomes diseased, sometimes even when the worm looks still healthy.

Epidermal lesions frequently appear in various places (fig. 47A–D). The most typical and serious lesions can be observed in *D. lugubris* on the dorsal side. First brown spots appear where the body wall has flattened out. Next the epidermis opens at these points and parenchyma escapes. If lesions appear on both sides, they are often symmetrical. Usually they follow each other, in a cephalocaudal direction, at regular intervals. In certain cases, the lesions appear as narrow slits parallel to the lateral edges and sometimes extend to the entire length of the animal. Healing cannot take place but the muscles contract strongly. Sooner or later the muscles relax and the lesions look like large, gaping openings that let necrotized tissue pass through (fig. 47E). The ventral epidermis can also be affected, but it does not yield to inner pressure. Some individuals keep the lesions throughout, partially or completely, only to reappear a few days later, in the same place or elsewhere. This temporary healing can be repeated.

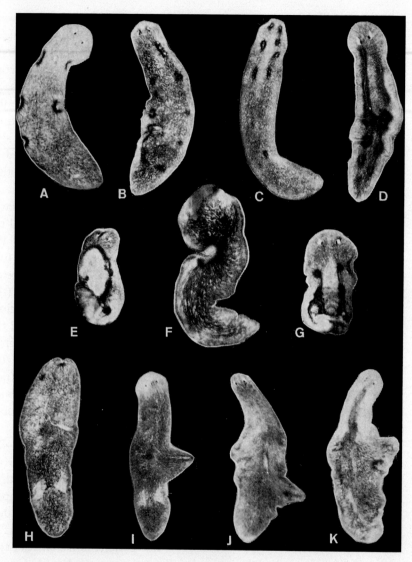

Fig. 47. The morphological symptoms of the disease in *D. lugubris. A* Marginal epidermic lesions. *B* Lateral lesions. *C* Lateral lesions a pair of which is above the eyes. *D* Lesions forming two longitudinal slits. *E* Opened lesion. *F* Marginal lesion formed after a tumor, head is regressed. *G* Regression of the tail. *H* Regression of the head, the two posterior white spots are healed lesions. *I* The same 11 days later, the head is regenerated and a lateral outgrowth formed at the level of the wound. *J, K* Formation of lateral excrescences after recuperation of marginal lesions, both have a differentiated head on their right side.

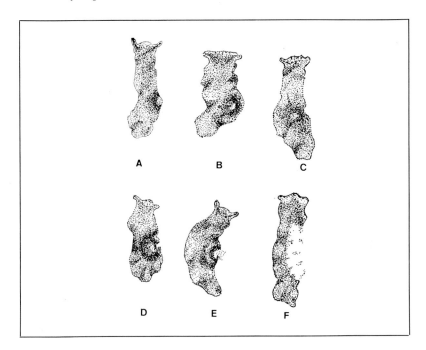

Fig. 48. Development of a postpharyngeal tumor in *P. cornuta. A–C* Three steps of tumor growth. *D* Necrosis begins in tissues adjacent to the tumor. *E* Disruption of epidermis; parenchyma is discharged with mucus. *F* The last stage before the death of the worm; histolysis progresses in all directions. From MARTELLI and CHANDEBOIS [1973].

The pigmentation often becomes abnormal. In many cases, this is actually the first symptom of the disease, almost inevitably preceding cephalic regression. The pigment of the head can change its color or its distribution, or disappears altogether.

In different groups of wounded planarians, malignant tumors were occasionally observed but the frequency never exceeded 10 %. On eyes and on auricules the tumors look like warts that grow quickly. In the pharyngeal and postpharyngeal zones they look like localized, lateral swellings. At the level of these tumors, the epidermis soon bursts. Healing does not take place and the surrounding tissues are progressively disintegrating (fig. 48).

The planarians usually exhibit sooner or later some very sudden and very violent contractions. When the head begins to regress, these contractions start abruptly whenever the worm brushes an obstacle. In later stages, a few hours before dying (when the body is swollen because of retention of necrotized internal tissues) the contractions recur for no apparent reasons almost as if the animal were in pain.

A histological study of sick worms is very difficult because the tissues are extremely fragile (the parenchyma is almost liquid) and they are deteriorated by fixation. Only a few observations were made so far with individuals bearing lesions or having regressed head. On histological sections, anomalies were only observed in the epidermis whose cells were missing here and there. Moreover, the epidermis forms pockets invaginating in the parenchyma. LANGE [1966] has already described such formations for planarians with tumors. However, they are not specific for this illness or tumor-bearing specimens for they can also be found in heads which have regenerated on contracted amputation surfaces [CHANDEBOIS, 1957a]. It is probably caused by the tissue reduction brought on by the disease. This reduction is uneven (slower in certain regions of the epidermis) and this may produce the inpocketings. On smears, one finds a large number of dead cells, but no type I cells or mitoses. The parenchyma looks like that of a starved planarian. This condition has been confirmed by electron microscopy. The nuclei are rounded, the endoplasmic reticulum is reduced, the intercellular spaces are clearly enlarged.

Most animals showing the symptoms of the disease never recover. However, a very few individuals do heal quite spontaneously and suddenly. They recover in less that 24 h, even though they are seemingly near to death. Their normal pigmentation returns and the epidermal lesions disappear. If the head regressed, a blastema forms and the animal completely regenerates (fig. 47I). However, in *D. lugubris,* this recovery is actually detrimental to individuals that had severe necroses. Abnormal growths develop rapidly, always at the expense of differentiated tissues and never through a blastema (fig. 47I–K). These growths are sometimes clearly distinguishable heads and tails, sometimes nonidentifiable end pieces or tubulous outgrowths that look like the 'benign' tumors described by several authors. They are located where the epidermal lesions have healed. When the latter are numerous and in a serial position, numerous regenerates will develop laterally so that the animal's contour appears serrated, even when it is completely stretched. It is probable that this disease occurs in nature too, because individuals were found with such supernumerary formations. TAR and TÖRÖK [1964] reported the spontaneous occurrence of double or multiple end pieces in *D. tigrina.* They attributed this to an over development of the central nervous system. However, it seems that these 'somatic twins' are, in fact, worms that have survived the disease described here. Such specimens are found in races which reproduce asexually (thus mutilate themselves spontaneously) and that have a tendency to produce benign or malignant tumors. STÉPHAN [1969] also described in the same species cases of serial supernumerary regenerates which gives the worm a serrated contour. He described inside these growths cavities whose walls are covered by the ventral epidermis and nerve cords connected to those of the host. The metamorphosis of a necrotized worm into a deformed monster, the price they pay for recovery, can be easily explained. In the diseased condition, the epidermis is completely disorganized; it becomes invaginated [MARTELLI and CHANDEBOIS, 1973] and breaks up. When recovery is taking place, the dorsal and ventral epidermis can join at certain points thus initiating the development of a distal end, either a head or a tail one (p. 123). The resulting intercalary regeneration is carried out entirely through morphallaxis, probably because of the partially dedifferentiated state of the tissues.

These observations lead us to a precise statement of the difference between 'malignant' and 'benign' or 'organized' tumors described by various authors. The malignant tumor is one of the *symptoms* of a disease; the benign tumor, when it is spontaneous, represents the *aftereffects,* caused by a local upheaval in the disposition of differentiated cells. This explains why the benign tumor can be obtained experimentally by simple mechanical means, such as thermocauterization [GOLDSMITH, 1940; STÉPHAN and SCHILT, 1966] or by insertion of inert bodies [SEILERN-ASPANG and KRATOCHWIL, 1965]. The benign tumor is organized and eventually develops eyes and auricles. The malignant tumor has the appearance of a simple swelling that apparently has no specific histological characteristics. FORSTER's experiments [1969] show that both types of tumors can appear in the same animal, side by side.

Though there are few symptoms for this disease, the syndrome and the course of events differ widely from one species to another, from one batch to another within the same species and even from one individual to another within the same batch. Up till now, *D. lugubris* has provided the most varied results. Within the same batch, there is always a large number of individuals that have only epidermic lesions, but in various places. Others undergo regression of the head, which can either be very quick or take several days. Head regression may be the only symptom of the disease, but usually it is preceded or followed by the appearance of epidermic lesions. The tumors appear sporadically. Some worms do not change in size, others are reduced conspicuously and become quite small. Some acquire granular black pigment; in others, the color fades away completely. In many cases, death follows the appearance of symptoms immediately, but in a few cases stricken specimens vegetate indefinitely, either as a complete organism deformed by swelling and lesions, or as a small fragment that cannot regenerate. Some specimens seem to recover spontaneously then have a relapse. These extremely varied reactions to a simple wound stand in contrast with the constancy of regeneration phenomena that are practically unchanged by the individual physiological variations (p. 114).

The parts lost by planarians during the disease, either after bursting of a tumor or by a cephalic regression, are reconstituted only in the very few cases which recover spontaneously. The loss of the regenerative faculties is obviously related to the disease's development. This loss is not caused by the lack of healing or failure of distalization, for the wound closure over the lesions looks perfectly normal. This was experimentally ascertained. If a part with lesions is cut out in a planarian which is only slightly affected, no blastema will form in most cases and the symptoms of the disease will manifest themselves again in the remaining parts. When a number of planarians under stress are amputated simultaneously (though they may look healthy) the fragments of the same batch will not behave all in the same way. Some of them will not regenerate, even though they bear no trace of lesions or pigmentary anomalies afterwards. Some regenerate normally. In other cases, by far the most frequent and interesting, regeneration begins in the usual manner. But it stops in various early stages, sometimes before differentiation of the eyes, more often before the formation of auricles (fig. 49 II, B, C). Morphallaxis does not take place, so that the pharynx is not reconstructed. This inhibition suggests that the operation was carried out just before the disease started to affect the course of regeneration. It begins, but is confined to an epimorphosis that is often unfinished. Some of these incomplete regenerates still look healthy. Others, especially the heads, regress rather rapidly, sometimes asymmetrically. Regression can be usually predicted, for the regenerates accumulate a bright Prussian blue pigment at an early stage when the controls are still colorless. In a great many cases, especially in *D. lugubris,* epidermal lesions also appear on the regenerate itself, above all around the eyes and in the mediodorsal region of the tails. When recovery is possible, the regenerate is reestablished (or completes its development). In fragments bound-

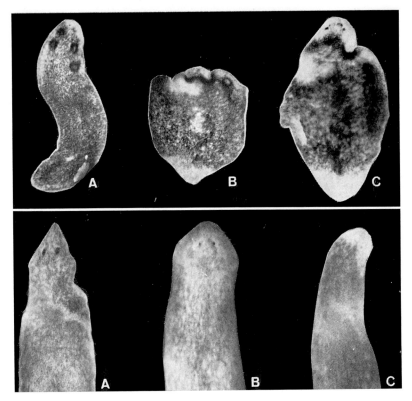

Fig. 49. Regeneration in sick planarians. *I D. lugubris. A* a 13-day-old regenerate at the beginning of the regression; *B* a prepharyngeal fragment regenerating since 10 days, the tail is normal, the head regressing, the dark pigment pushed backwards; *C* the same fragment 2 days later, the head is regenerated but the concentrated dark pigment is still visible at the base of the regenerate. *II. D. subtentaculata. A* head regression in a worm that has been wounded in the prepharyngeal zone; *B* normal 8-day-old regenerate in a small fragment isolated 3 days after wounding. *C* blocked 8-day-old regenerate in a worm wounded 3 days before decapitation.

ed by two amputation surfaces, the two regenerates often develop very differently. For example, in a planarian *(D. lugubris)* whose head and postpharyngeal zone were amputated, regeneration occurred at both ends, but then cephalic regression took place while the tail remained intact (fig. 49 I, B, C). In the stump the pharynx was expelled and abnormal pigmentation was observed. The whole individual behaved like a complete planarian whose head and pharynx regress while the posterior extremity remains.

Statistical results indicate that in a single batch, the distribution of anomalies is the same in sick animals that were not amputated and in those that regenerated. This clearly shows that the disease appeared in the regenerated parts approximately in the same way as in the corresponding parts of the animal left unamputated.

The most striking aspect of the disease is the close resemblance between its symptoms and the consequences of X-ray treatment (p. 67). In uncut organisms, the spatiotemporal patterns for regressions and epidermic lesions are the same. Death occurs when the swelling of the body causes bursts of the epidermis at the level of the lesions and expulsion of necrotized internal tissues. In both the spontaneous and the irradiation-induced disease, the affected animals show involution of the type I cell system and breaks in the dorsal epidermis. The spontaneous later recovery of the regenerative faculty resembles recovery after a nonlethal dose of irradiation. When regeneration takes place just before the disappearance of the regenerative faculty, it is incomplete, in both cases. Epimorphosis remains unfinished. For example, individuals that rebuild auricles are rare. There is no morphallaxis, so that the pharynx cannot be reconstructed. The sick regenerates often regress like regenerates of irradiated worms, the only difference being the distinctive color of the pigment that they accumulate. It is thus obvious that the planarian disease described here consists of a blocking of the differentiation process which is manifested by inhibition of cell renewal (lesions) and by an apparently abrupt loss of the regenerative faculties. This does not eliminate other disorders revealed by various pigmentary anomalies and a generalized swelling of the body. The constancy of the patterns of the lesions and regressions is probably related to the constancy of cell renewal patterns, but their dynamics are unfortunately still unknown to us.

Before attempting to explain the causative factors of the disease, one must first make sure that it was not caused by microorganisms or by toxic substances. For the purpose of verification, the worms were operated under aseptic conditions, treated with antibiotics, kept in Pyrex dishes cleaned as for tissue culture. Postoperative mortality decreased markedly, but the disease manifested itself exactly as in batches manipulated under normal experimental conditions. Pollution also had to be excluded as a factor that predispose the disease. *D. subtentaculata* and *P. cornuta* were collected in mountain springs far from cities or cultivated fields upstream.

The disease is not the consequence of the injury of a specific organ or body part because the localization and the type of the wound do not

affect in the least the course of the disease. MARTELLI [1970] used pre-
and postpharyngeal injections; the present author also used various kinds
of interventions: making an incision in the dorsal epidermis, eliminating a
narrow band of tissues around the tail's edge, making a transversal incision in front of or behind the pharynx. In all the cases, the lesions appear
independently from the level of the wound and head regression occurs
only in a certain number of cases.

Since the disease can be brought on experimentally by wounding individuals one must look for the causes of the disease among the immediate
consequence of a trauma without tissue loss. These consequences were already discussed. The mitotic activity of the type I cell system is stimulated
throughout the parenchyma as a result of water penetration (p. 62). Sub-
sequently, substances are released (everything seems to point to nucleic
acids, especially RNA). Several days later they can still be used by differentiating cells (p. 73). In the regenerating animal, the surplus is used up
naturally for the formation of the blastema. It seemed therefore possible
that the reason for the wounded animal's diseased condition can be found
in the excess of substances that new cells around the scar cannot resorb
entirely. Obviously this is just a working hypothesis, but there is already
indirect proof to support it.

In experimental and natural conditions that stimulate the activities of
the type I cell system, the disease strikes all the wounded worms. On the
contrary, high percentages (80 % in some batches) of worms remain
healthy after wounding, if they were previously starved for 1–3 months,
or when they were collected at the end of winter. An important fact is the
greater resistance of starved planarians to wounding and to X-rays alike.
In exceptional cases food alone can be a cause of the disease. In 1966,
LANGE described the appearance of malignant tumors in batches of *D.
etrusca* fed twice a week instead of once. The tumors necrotized rapidly;
the tissues disintegrated progressively and the animal died in the end.
On the basis of this experimental study, which we did not repeat, one
can say that planarians may sometimes develop the disease without previous wounding, when their diet is too rich. This could explain why in
many cases the disease can manifest itself without previous wounding. In
numerous experiments we did, in fact, observe lesions and regressions in
controls, the proportions varying according to the batches. Such symptoms in controls were especially frequent in *D. lugubris* collected at the
beginning of September and operated immediately. At first the phenomenon was explained by assuming accidental lesions during collection,

especially in *D. lugubris* which stick closely to the substrate (pebbles) at certain points of the body almost as if they had pseudopodia. However, in spite of all the necessary precautions to avoid such accidental lesions, the disease still appeared spontaneously in the controls and sometimes in such large numbers that we were inclined to look for other causes.

Extensive regeneration before wounding also preserves the animals from the disease and even saves animals that are already under stress. For example, in a batch of *D. subtentaculata* both controls and wounded worms showed lesions and died at high rate. Nevertheless, prepharyngeal fragments isolated 3 days after the worms were collected regenerated completely within the normal period of time. Their recovery was considered as complete since wounds made 2 weeks after the amputations did not produce disease nor prevent a second complete regeneration.

All this preliminary research enabled us to put forward a tentative explanation of the cause and general course of the planarian disease (fig. 50). The purpose of this explanation is to present a framework for further research in an entirely unexplored domain, that is related to very pertinent up-to-date problems of regeneration and related phenomena. In the animal predisposed to developing the disease, the type I cell system is very active. The nucleic acids freed by cytolysis are utilized in the production

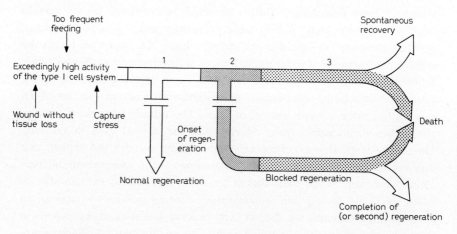

Fig. 50. Cause and progress of the planarian disease reported here. 1 = Latent period, the illness will not manifest itself if regeneration occurs; 2 = the diseased condition is irreversible, the worms amputated during this period regenerate partially; 3 = last period, regeneration is impossible.

of differentiated cells (cell renewals and growth). The type I cell system and the other tissues are normally in a state of equilibrium. Any factor that can stimulate the mitotic activity of the type I cell system upsets this balance, because it releases excess nucleic acids. Excessive feeding probably leads to a maximum molecular and cellular turnover, while the activity of the type I cell system continues to increase. The rapid stimulation of the type I cell system after wounding would provoke a similar massive production of nucleic acids over and above of what is required for the actual needs of differentiation and growth. Stress brought on by trauma suffered in the course of the collection of the specimens from their natural habitat could be explained by an abrupt modification of neurosecretions which could also provoke activation of the type I cell system. This interpretation was reinforced by the fact that in amphibians simple handling of breeding animals is known to produce significant changes in the cell composition of the blood. In all these cases, the excess nucleic acids upset certain specific metabolic processes. One could envisage these excess nucleic acids as fuel that 'floods' the motor car engine. In insufficiently fed animals, the molecular and cellular turnover is reduced, as proven by the reduction in size and decrease in the cell number (p. 7). It can increase again when the food supply returns to normal. Thus the organism is ready to handle an increased production of nucleic acids. The resistance of just regenerated fragments to the disease can be similarly explained, since the regenerative growth is not finished and there is no food intake from the beginning of the experiment. 'Cinephyllaxis' (p. 63) must not be excluded, either, since a wound inflicted a week after the first amputation cannot stimulate the type I cell system a second time. Thus, starvation gives the animal a certain 'margin for safety' in case of the system's sudden activation, while the cinephyllaxis prevents 'flooding' which would certainly occur after repeated lesions without this 'safety valve'. However, the well-fed worm is constantly on the edge of the danger, and if it is overfed, it shortens its own life.

When an excess of nucleic acids appears in an animal, the situation is not immediately irreversible. Regeneration can save it. Thus the disease's first phase is a latent period. Once the disease is irreversibly determined two phases follow: the second phase, during which production of differentiated cells is still possible, and the third, preceding either death or recovery, when differentiation is no longer possible. If amputation is made during the second phase, regeneration begins, but has no time to be completed before the start of phase three. The proximal levels will not be formed and in those parts which have been rebuilt, cell differentiation will not be completed. The disease's symptoms will appear in the blocked regenerate, as well as in the corresponding parts of the body of a worm which was not amputated and is in the same stage of the disease. If there is a recovery, regeneration is completed or a second one takes place to replace the eliminated parts.

Certain seriously affected animals with numerous lesions can sometimes survive for a long time; others in which only the heads have regressed die right away. These simple observations show that the disease's lethal character is not the direct and exclusive consequence of an inhibited differentiation, but rather the result of indirect effects that have repercussions on several systems. Depigmentation, or change in pigmentation (the Prussian blue color of regenerates which start to regress), generalized swelling of the body, involution of the type I cell system are all signs of a general metabolic imbalance, whose severity is not related to that of the visible lesions. The individual variations in the organism's fight against the disease also indicates that the physiological condition plays a role; normally fed worms, though the most likely to develop the disease, are also the most likely to show spontaneous recovery.

Despite the rather specific symptoms, the planarian disease does not seem to be different from cancer. If there are no tumors formed as in mammals, this can be explained by the aberrant nature of the undifferentiated cells, whose ultrastructure was discussed in detail in earlier chapters. Let us suppose that a tumor develops locally in the parenchyma. According to our concept, cells will return to (or stay in) an undifferentiated state, that is, the syncytial condition. As their number increases, they will form a liquid pocket, containing almost exclusively nuclei. As they progressively become more abundant, the pocket expands and the internal pressure increases. If the epidermis enclosing the tumor bursts, the pocket is suddenly emptied, leaving in its place an opening that spreads as the disease progresses. This is how 'malignant tumors', as they are called by all authors, develop in planarians. Electron-microscopic observations on regenerating worms have shown that during a major dedifferentiation (for example, during regeneration) the syncytium infiltrates between the parenchymal cells and separates them from each other. Let us suppose that this phenomenon occurs in the cancerous state: a visible tumor will not appear. The phenomenon of syncytium formation was observed with electron microscopy, but one cannot exclude the possibility that dedifferentiation is a secondary effect of a disease related to poor physiological conditions, analogous to those of an individual starved for several months. Along with the inability of the syncytium to develop, one must also think of the local disappearance of differentiated cells in cancer. This is probably the cause of epidermal lesions and regressions as indicated by their similarity to the effects of X-ray treatment.

ERTL [1973] underlined the very important fact that a malignant tumor in mammals is not lethal in itself. It causes functional insufficiencies and the dysfunction of certain organs and these 'systemic effects' will finally lead to the death of mammals with tumors. Among these effects ERTL noted the complete involution of the thymus, the delay of liver regeneration after partial hepatectomy, the progressive decrease of the net weight of the body and symptoms of cachexia. Another common point between the planarian disease and cancer in mammals is that the disease is not a simple local failure of a few cells to integrate their activity with that of adjacent tissues. In the terminal phase of the planarian disease the entire worm suffers from metabolic imbalance, as we have seen. There are some other striking resemblances to mammalian cancer. In particular, the involution of the type I cell system resembles the involution of the thymus; the loss of the regenerative faculty, the delayed regeneration of the liver; the reduction of the size, the symptoms of cachexia.

Whether a tumor is malignant can only be determined after it has been transplanted onto a healthy individual, where it should produce metastases. In planarians this test is technically impossible since the sick individual's parenchyma is liquid. Thus, there is little hope that one day one may use these theoretically very interesting worms as standard material for experimental cancer research. But the analogy between the type I cell system and the vertebrate thymus remains a fact and one still has to pose questions. Do the enormous quantities of thymocytes, destroyed by the organism every day, really have the trophic role that some authors have suspected for more than 50 years? If so, could they, when in excess, play a pathological role in the organisms, like the type I cells in planarians and determine the appearance of malignant tumors? During the last few years the involution of the thymus and its role in the immunological function of the cancerous individuals has been studied to a great extent. But the hyperfunction of this organ has so far not been suspected as a factor that could have a role in carcinogenesis. It is possible that civilized man, like the overfed planarian, lives without that 'margin of safety' that prevents the catastrophic effects of an external factor stimulating the thymus. Unfortunately, among these external factors, certain therapeutic methods

– whose deferred and possibly carcinogenic effects are difficult to control – could be the most deadly. Perhaps there is a certain *'ars vivendi'* that could prevent cancer? It would be a rather chastising thought if we would learn it from the lowly planarians!

Epilogue

At the present time ultrastructure and molecules captivate the interest of biologists, to such an extent that research with cells and organs is often considered improductive, like working in an exhausted gold mine. Not only is the approach above the molecular level deemed worthless, but even the worker is discredited. The author of this monograph, as she is about to sum up the conclusions of her work, emphasizing the need to formulate concepts based on cell sociology and a cautious attitude to the premature use of the latest techniques on planarians, she fully realizes that this will draw sharp criticism.

To be better prepared for the anticipated attack it may be profitable to cite the example of space research here. The success of this adventure is due not only to the construction of rockets and equipment that conveys information about the cosmos to earth. If man had not acquired an extremely accurate idea of the universe, through centuries of exploration, observation and calculation, the rockets might have been lost in space or else the data collected could not have been processed. Biochemical research with organism as complex as planarians can be compared to space research. The perfection of techniques used to probe the molecular universe, inaccessible to our senses, is not the only guarantee for success. One must also have a thorough knowledge of the targets that are aimed at and how to probe them in the right direction. One must be capable of interpreting the resulting information. This will be impossible as long as we do not know how cells interact with each other, how they renew and destroy themselves in the healthy as well as in the regenerating organism.

Belief in concepts accepted as final and unchangeable has led experimenters to reducing planarians to a far too simple scheme. In this way the regenerating worm becomes simply a bag of 'neoblasts' since these cells are all what is needed for repairing tissues and replace amputated organs. Similarly, if one sticks to the idea that the planarian head dominates the rest of the structure of the body, there is reason for reducing the entire organization to four zones, with the eyes, the pharynx and the copulatory system as the only morphological references. Once these simplifications are accepted, there is nothing to prevent the study of planarians with the same methods used for bacterial cultures or presumptive areas of young embryos. By assuming gradients in the concentration or activity of certain molecules, one can claim that certain properties of 'neoblasts' have been found. If an experimental factor blocks regeneration, this can be explained by the assumption that it has destroyed or inhibited the 'neoblasts'. If one extracts a substance from an organ that prevents regeneration of the same organ, it is the simplest explanation to conclude that this substance has inhibited identical specific synthesis. However, a closer analysis usually reveals that the situation is more complicated than assumed and the conclusion will not explain all phenomena. If one is reluctant to demolish accepted theories, there are only two ways open: either to avoid the discussion of the basic questions or to look for an interpretation which, even if very unlikely, agrees with the existing concept. One of the most symptomatic examples is the case of acid phosphatases (discussed on p. 29). Their increased activity at the beginning of regeneration has been attributed to cell dedifferentiation. AUTUORI et al. [1965] maintained this interpretation for annelids where this phenomenon is now accepted; in planarians, acid phosphatase activity was related to nutritional functions because dedifferentiation would be incompatible with the notion of 'neoblasts' whose existence the authors did not want to dispute. It is neither a negative attitude nor a desire to revive quarrels that makes us to point out the futility of using the latest techniques of molecular biology on living organisms of which we still have imperfect conceptions: the application of new techniques will not correct old mistakes and may even produce new ones. If this monograph can only convince developmental biologists of the need to revise classical concepts of histogenesis and morphogenesis in planarians, it has fulfilled its most important purpose.

As for the new biochemical methods, let us take a look at them in the light of the main conclusions of this monograph. These conclusions were that (1) an amputation triggers both the dedifferentiation of certain cells and the stimulation of a system functionally analogous to the thymus; (2) regeneration is programmed during wound healing and implies epigenetic phenomena. The results given by the biochemical methods are not clear cut and open to different interpretations. Has a given factor that inhibits blastema formation interfered with the metabolism of dedifferentiated cells? Or prevented the programming of regeneration? Or blocked the developmental clock (for example by making inductive properties appear before the competent state is reached in cells that will be induced)? If one can speed up the restitution of eyes, is it by speeding up differentiation or by allowing the animal to have at its disposal cells activated sooner after amputation? When one analyzes the effects of a substance during various stages of regeneration, one does not necessarily study effects on the activity of cells that build up the blastema, because the effect on regeneration as a whole is the sum of the effects on the activity of all the systems involved in regeneration. If, for example, the amount of DNA in crushed planarians is measured, the variations recorded are those of the syncytium and of the type I cells. Since mitoses in the latter are much more numerous, the curve shows the symptoms of an inflammatory reaction and not the activity of elements directly involved in histogenesis. Evaluation of the thymidine kinase activity [COWARD *et al.*, 1970] is a more precise and elegant method than DNA determination. It allowed COWARD *et al.* to suspect the existence of two distinct cell activities during regeneration, but not to demonstrate their existence nor to pinpoint their role in regeneration (p. 64). Likewise, when regenerating planarians are treated with a substance whose effect on a specific metabolic pathway or on a cell function is well-known, one does not know what the target of this substance is. It is never certain whether the effects produced are due to a poisoning of certain tissues or to some other mechanism. The example of mitoclastic poisons is one of the most striking cases (p. 130). During prolonged treatments, these substances inhibit regeneration even though regeneration is possible without mitoses. When applied within a few hours after amputation, mitoclastic substances may produce heteromorphic regeneration which in other circumstances would be accompanied by considerable mitotic activity. If one adds to the list of the factors that affect regeneration the metabolic disease, caused by the most harmless circumstances, one understands even better the hazards of the premature use of certain techniques on material that is not yet fully understood.

In spite of all these uncertainties we should not reach 'defeatist' conclusions. One cannot have an accurate idea of the mechanics of a machine by simply provoking breakdowns, without knowing how the machine works. First one has to take it apart to understand the construction and the roles of the components; only then can one explain the breakdowns. Likewise we should first study the activity of various types of cells that make up the planarian body and how these cellular activities are coordinated to assure regeneration. A good number of results whose ambiguity is disappointing could finally be interpreted and brought into a coherent picture. The new concepts presented in this monograph were developed largely on the basis of using histological preparations. The concepts are not intended to be the articles of a new *credo,* to be imposed upon all who study planarian regeneration. They have to be examined further and tested from new angles to be either confirmed or revised and modified, even repeatedly, if necessary. The thousands of publications on planarian regeneration which have already appeared in print should not delude us: there is still much to be done and much needs to be cleared up about the real problems of morphogenesis in these organisms.

References

ABELOOS, M.: Recherches expérimentales sur la croissance et la régénération chez les Planaires. Bull. biol. Fr. Belg. *64:* 1–140 (1930).

ABELOOS, M.: La régénération et les problèmes de la morphogenèse (Gauthier-Villars, Paris 1932).

ABELOOS, M.: Régénération postérieure chez *Magalia perarmata*. C. r. hebd. Séanc. Acad. Sci., Paris *230:* 447–478 (1950).

ABELOOS, M.: L'histogenèse dans la régénération des Vers; in KIORTSIS and TRAMPUSCH Regeneration in Animals, pp. 216–227 (North-Holland, Amsterdam 1965).

ABELOOS, M. et LECAMP, M.: Sur la taille des cellules épithéliales tégumentaires au cours de la croissance et au cours du jeûne chez les Planaires. C. r. Séanc. Soc. Biol. *101:* 899–901 (1929).

ANSEVIN, K. D.: The influence of a head graft on regeneration of the isolated post-pharyngeal body section of *Dugesia tigrina*. J. exp. Zool. *171:* 235–242 (1969).

ANSEVIN, K. D. and BUCHSBAUM, R.: Observations on Planarian cells cultivated in solid and liquid media. J. exp. Zool. *146:* 153–161 (1961).

AUTUORI, F.; BUONGIORNO-NARDELLI, M. et GABRIEL, A.: Les enzymes protéolytiques et les phosphatases au cours de la régénération de fragments de Planaires. C. r. hebd. Séanc. Acad. Sci., Paris *260:* 995–998 (1965).

AUTUORI, F.; BUONGIORNO-NARDELLI, M. et THOUVENY, Y.: Activités enzymatiques dans la régénération de l'Annélide *Hydroïdes norvegica*. C. r. hebd. Séanc. Acad. Sci., Paris *260:* 1274–1276 (1965).

AVEL, M.: Sur une expérience permettant d'obtenir la régénération d'une tête en l'absence certaine de la chaîne nerveuse ancienne chez les Lombriciens. C. r. hebd. Séanc. Acad. Sci., Paris *194:* 2166 (1932).

BANDIER, J.: Histologische Untersuchungen über die Regeneration von Landplanarien. Roux Arch. EntwMech. Org. *135:* 316–348 (1936).

BARDEEN, C. R.: Embryonic and regenerative development in Planarians. Biol. Bull. *3:* 262–288 (1902).

BARDEEN, C. R. and BAETJER, F. H.: The inhibiting action of the roentgen rays on regeneration in Planarians. J. exp. Zool. *1:* 191–195 (1904).

BART, A.: Conditions locales du déclenchement et du développement de la régénération d'une patte chez l'Insecte *Carausius morosus*. C. r. hebd. Séanc. Acad. Sci., Paris *299:* 473–476 (1969).

BARTSCH, O.: Die Histogenese der Planarianregenerate. Roux Arch. EntwMech. Org. *99:* 187–221 (1923).

BEAUCHAMP, P. DE: Classe des Turbellariés; in GRASSÉ Traité de zoologie, vol. 4, pp. 69–123 (Masson, Paris 1961).

BEST, J. B.; HAND, S., and ROSENVELD, R.: Mitosis in normal and in regenerating Planarians. J. exp. Zool. *168:* 157–168 (1968).

BETCHAKU, T.: Isolation of Planarian neoblasts and their behaviour *in vitro* with some aspects of the mechanism of the formation of regeneration blastema. J. exp. Zool. *164:* 407–434 (1967).

BETCHAKU, T.: The cellular mechanisms in the formation of a regeneration blastema of the fresh-water Planaria. I. The behaviour of cells in a tiny body fragment isolated *in vitro.* J. exp. Zool. *174:* 253–280 (1970).

BOHN, H.: Analyse der Regenerationsfähigkeit der Insektenextremität durch Amputations- und Transplantationsversuche an Larven der afrikanischen Schabe *(Leucophaea madera).* Roux Arch. EntwMech. Org. *156:* 449–503 (1965).

BOILLY, B. et BOILLY-MARER, Y.: Rôle des contacts tissulaires dans le déclenchement de la morphogenèse régénératrice chez la Nereis. Bull. Soc. zool. Fr. *97:* 295–310 (1972).

BRØNDSTED, H. V.: Regeneration in planarians investigated with a new transplantation technique. Kgl. D. Vid. Selsk. Biol. Medd. *15:* 1–39 (1939).

BRØNDSTED, H. V.: Further experiments on regeneration problems in planarians. Kgl. D. Vid. Selsk. Biol. Medd. *17:* 1–27 (1942).

BRØNDSTED, H. V.: The existence of a static, potential and graded regeneration field in planarians. Kgl. D. Vid. Selsk. Biol. Medd. *20:* 1–31 (1946).

BRØNDSTED, H. V.: Planarian regeneration. Biol. Rev. *30:* 65–126 (1955).

BRØNDSTED, H. V.: Planarian regeneration. Int. Ser. Monogr. in Pure and Applied Biol. Zool. Div., vol. 42 (Pergamon Press, Oxford 1969).

BRYANT, B. J.: Réutilisation de l'ADN des leucoytes par des cellules du foie en régénération. Expl. Cell Res. *27:* 70–79 (1962).

BUCHANAN, J. W.: Regeneration in *Phagocata gracilis* (Leidy). Physiol. Zool. *6:* 185–204 (1933).

BULLIÈRE, D.: Utilisation de la régénération intercalaire pour l'étude de la détermination cellulaire au cours de la morphogenèse chez *Blabera craniifer* (Insecte, Dictyoptère). Devl Biol. *25:* 672–709 (1971).

BURNETT, A. L.: Control of polarity and cell differentiation through auto inhibition. Expl. Biol. Med., vol. 1, pp. 125–140 (Karger, Basel 1967).

CARLSON, B. M.: The regeneration of minced muscles. Monogr. devl Biol., vol. 4 (Karger, Basel 1972).

CASTLE, W. A.: The life history of *Planaria velata.* Biol. Bull. *53:* 139–144 (1927).

CASTLE, W. A.: An experimental and histological study of the life cycle of *Planaria velata.* J. exp. Zool. *51:* 417–483 (1928).

CECERE, F.; GRASSO, M.; URBANI, E. e VANNINI, E.: Osservazioni autoradiografiche sulla rigenerazione di *Dugesia lugubris.* Rend. 1st. Sci. Camerino. *5:* 193–198 (1964).

CHANDEBOIS, R.: Inhibition partielle de la régénération chez la Planaire marine *Procerodes lobata.* C. r. hebd. Séanc. Acad. Sci., Paris *231:* 1347–1348 (1950).

CHANDEBOIS, R.: Etude expérimentale des régénérations asymétriques chez la Planaire marine *Procerodes lobata.* Bull. Soc. zool. Fr. *76:* 404–408 (1951).

CHANDEBOIS, R.: Hétéromorphoses et hémihétéromorphoses chez la Planaire marine *Procerodes lobata*. C. r. hebd. Séanc. Acad. Sci., Paris *234:* 1319–1321 (1952).

CHANDEBOIS, R.: Sur le phénomène d'hémihétéromorphose chez la Planaire marine *Procerodes lobata*. Bull. Soc. zool. Fr. *80:* 139–144 (1953).

CHANDEBOIS, R.: Recherches expérimentales sur la régénération de la Planaire marine *Procerodes lobata*. Bull. biol. Fr. Belg. *91:* 1–94 (1957a).

CHANDEBOIS, R.: Détermination de la région pharyngienne chez la Planaire marine *Procerodes lobata*. C. r. hebd. Séanc. Acad. Sci., Paris *245:* 1177–1178 (1957b).

CHANDEBOIS, R.: Sur l'«induction» du pharynx dans la régénération des Planaires. Bull. Soc. zool. Fr. *84:* 434–439 (1959).

CHANDEBOIS, R.: Sur la source de l'histogenèse régénératrice chez les Planaires. C. r. hebd. Séanc. Acad. Sci., Paris *251:* 146–148 (1960).

CHANDEBOIS, R.: Rôle des éléments fixes et libres du parenchyme dans la régénération de *Planaria subtentaculata*. Bull. biol. Fr. Belg. *96:* 203–227 (1962).

CHANDEBOIS, R.: Cultures de fragments de Planaires d'eau douce. Ann. Epiphyties *14:* 141–152 (1963a).

CHANDEBOIS, R.: Action des rayons X sur la différenciation cellulaire de *Planaria subtentaculata*. Bull. Soc. zool. Fr. *86:* 632–644 (1963b).

CHANDEBOIS, R.: Cell transformation systems in Planarians; in KIORTSIS and TRAMPUSCH Regeneration in Animals, pp. 131–142 (North-Holland, Amsterdam 1965a).

CHANDEBOIS, R.: Variations du pouvoir régénérateur de Planaires irradiées *(Dugesia subtentaculata)* en fonction de la dose de rayons X. C. r. hebd. Séanc. Acad. Sci., Paris *260:* 4834–4837 (1965b).

CHANDEBOIS, R.: Action d'un milieu synthétique convenant pour des cultures histiotypiques sur l'activité du tissu indifférencié de fragments de Planaires. 2nd Int. Coll. on Invertebrate Tissue Culture, Como 1967, pp. 32–62 (Fusi, Pavia 1968a).

CHANDEBOIS, R.: The respective roles of mitotic activity and of cell differentiation in Planarian regeneration. J. Embryol. exp. Morph. *20:* 175–188 (1968b).

CHANDEBOIS, R.: L'histogenèse chez les Planaires. Ann. Biol. *9:* 543–554 (1970).

CHANDEBOIS, R.: Augmentation du pouvoir de régénération chez des Planaires irradiées à doses létales par la stimulation du système des cellules de type I. C. r. hebd. Séanc. Acad. Sci., Paris *272:* 1417–1419 (1971a).

CHANDEBOIS, R.: Mise en évidence du phénomène de cinéphylaxie dans le système des cellules de type I de la Planaire *Dugesia subtentaculata*. C. r. Séanc. Soc. Biol. *169:* 1323–1326 (1971b).

CHANDEBOIS, R.: An analysis of mitotic activity and the question of a thymus-like system in Planarians. Oncology *26:* 540–566 (1972).

CHANDEBOIS, R.: The syncytial and intercellular nature of dedifferentiated material in the parenchyma of regenerating and starved Planarians. Oncology *27:* 356–384 (1973a).

CHANDEBOIS, R.: General mechanisms of regeneration as elucidated by experiments on Planarians and by a new formulation of the morphogenetic field concept. Acta biotheoretica *22:* 2–33 (1973b).

CHANDEBOIS, R.: Teneurs en ADN des éléments qui participent à la régénération chez la Planaire dulcicole *Dugesia subtentaculata.* C. r. Soc. Biol. Fr. *67:* 286–291 (1973c).

CHILD, C. M.: The asexual cycle of *Planaria velata* in relation to senescence and rejuvenescence. Biol. Bull. *25:* 187–203 (1913).

CHILD, C. M.: Asexual breeding and prevention of senescence in *Planaria velata.* Biol. Bull. *26:* 286–293 (1914).

CHILD, C. M.: Patterns and problems of development (Chicago 1941).

CLARK, M. E. and CLARK, R. B.: Growth and regeneration in *Nephthys.* Zool. Jb. Physiol. *70:* 25–85 (1962).

COHEN, F.: Sur l'évolution des cellules embryonnaires du parenchyme des Planaires au cours du jeûne. C. r. Séanc. Soc. Biol. *131:* 1216–1218 (1939).

CORNEC, J. P. et FONTAINE, M.: Etude expérimentale des asymétries dans la différenciation des yeux observées au cours de la régénération chez *Polycelis cornuta* Johnson. C. r. hebd. Séanc. Acad. Sci., Paris *263:* 1411–1414 (1966).

COULOMB-GAY, R. M.: Hémihétéromorphoses chez le Lombricien *Eiseniella tetraedra f. typica.* C. r. hebd. Séanc. Acad. Sci., Paris *275:* 1633–1636 (1972).

COULOMB-GAY, R. M. et CORNEC, J. P.: Localisation des activités phosphatasiques après amputation chez les Annélides. C. r. Séanc. Soc. Biol. *167:* 1634–1641 (1973).

COWARD, S. J.; HIRSCH, F. M., and TAYLOR, J. H.: Thymidine kinase activity during regeneration in the planarian *Dugesia dorotocephala.* J. exp. Zool. *173:* 269–277 (1970).

CRÉMIEU, M.: Etude des effets produits sur le thymus par les rayons X; thèse de médecine Lyon (1912).

CURTIS, W. C.: The life history, the normal fission and the reproductive organs of *Planaria maculata.* Proc. Boston Soc. Nat. Hist. *30:* 515–559 (1902).

CURTIS, W. C. and HICKMAN, J.: Effects of X-rays and radium upon regeneration in Planarians. Anat. Rec. *34:* 145–146 (1926).

DE BOTH, N. J.: The developmental potencies of the regeneration blastema of the Axolotl limb. Roux Arch. EntwMech. Org. *165:* 242–276 (1970).

DECOSSE, J. J. and AIELLO, N.: Feulgen hydrolysis: effect of acid and temperature. J. Histochem. Cytochem. *14:* 601–604 (1966).

DESSELLE, J. C.: Restauration de la régénération par des implants de cartilage dans les membres irradiés de *Triturus cristatus.* C. r. hebd. Séanc. Acad. Sci., Paris *267:* 1642–1645 (1968).

DRESDEN, J. D.: Pharynx regeneration in *Polycelis nigra.* Acta neerl. Morph. Norm Path. *3:* 140–150 (1940).

DUBOIS, F.: Contribution à l'étude de la migration des cellules de régénération chez les Planaires dulcicoles. Bull. Biol. Fr. Belg. *83:* 213–283 (1949).

DUSTIN, A. P.: Les variations saisonnières du thymus de la grenouille. Bull. Soc. Roy. Sci. med. nat. (1912).

DUSTIN, A. P.: Recherches d'histologie normale et expérimentale sur le thymus des amphibiens anoures. II. Histogenèse normale et influence de l'alimentation sur l'histogenèse. Archs Biol., Paris *30* (1920).

DUSTIN, A. P.: Thymocytes and lymphocytes. Revue fr. Endocr. clin. *1:* 332–345 (1923).

DUSTIN, A. P. et GRÉGOIRE, C. H.: Contribution à l'étude de la mitose diminutive ou élassotique dans le thymus des Mammifères. C. r. Séanc. Soc. Biol. *108:* 1159 (1931).

ERTL, N.: 'Systemic effects' during the growth of malignant experimental tumors. Oncology *27:* 415–429 (1973).

FABER, J.: Autonomous morphogenetic activities of the amphibian regeneration blastema; in KIORTSIS and TRAMPUSCH Regeneration in Animals, pp. 404–419 (North-Holland, Amsterdam 1965).

FABER, J.: Vertebrate limb ontogeny and limb regeneration: morphogenetic parallels. Adv. Morphogenes. *9:* 127–147 (1971).

FEDECKA-BRUNER, F.: Régénération des testicules des Planaires après destruction par les rayons X; in KIORTSIS and TRAMPUSCH Regeneration in Animals, pp. 185–192 (North-Holland, Amsterdam 1965).

FLEXNER, S.: The regeneration of the system of *Planaria torva* and the anatomy of double-headed forms. J. Morph. Physiol. *14:* 337–346 (1898).

FLICKINGER, R. A.: A gradient of protein synthesis in *Planaria* and reversal of axial polarity of regenerates. Growth *23:* 251–271 (1959).

FLICKINGER, R. A.: Isotopic evidence for a local origin of blastema cells in regenerating Planaria. Expl. Cell Res. *34:* 403–406 (1964).

FORSTER, J. A.: Malformations and lethal growths in planaria treated with carcinogens. Natn. Cancer Inst. Monogr. *31:* 683–691 (1969).

FRANQUINET, R.: Cultures *in vitro* de cellules de la Planaire d'eau douce *Polycelis tenuis* Iijima. C. r. hebd. Séanc. Acad. Sci., Paris *276:* 1733–1736 (1973).

GABRIEL, A.: Etude morphologique et évolution biochimique des néoblastes au cours de la première phase de la régénération des Planaires d'eau douce. Ann. Embr. Morph. *3:* 49–70 (1970).

GABRIEL, A. et LE MOIGNE, A.: Action de l'actinomycine D sur la différenciation cellulaire au cours de la régénération de Planaires qui viennent d'éclore. I. Etudes morphologiques, histologiques et ultrastructurales du pouvoir de régénération en présence de l'antibiotique. Z. Zellforsch. mikrosk. Anat. *115:* 426–441 (1971a).

GABRIEL, A. et LE MOIGNE, A.: Evolution de l'incorporation de leucine tritiée dans le blastème et des taux de mitoses au cours de la régénération des jeunes planaires traitées par l'actinomycine D. C. r. hebd. Séanc. Acad. Sci., Paris *272:* 2017–2020 (1971b).

GAZSÓ, L. R.: Contribution à l'étude de la régénération du pharynx du *Dendrocoelum lacteum* et de la *Dugesia lugubris*. Act. Biol. Acad. Sci. Hungaricae *8:* 263–272 (1958).

GOLDSMITH, E. D.: Spontaneous outgrowths in *Dugesia tigrina*. Anat. Rec. *75:* suppl., pp. 158–159 (1940).

GOODWIN, B. C. and COHEN, M. H.: A phase-shift model for the spatial and temporal organization of developing systems. J. theor. Biol. *25:* 49–107 (1969).

GOSS, R. J.: Regenerative inhibition following limb amputation and immediate insertion into the body cavity. Anat. Rec. *126:* 15–27 (1956).

GOUTIER, R.: Effects of X-rays on nucleic acid biosynthesis and on the activity of nucleases in Mammalian cells. Prog. Biophys. *11:* 53–57 (1961).

GUYOT, M.; POUSSEL, H. et GAVAUDAN, P.: Comparaison de l'action exercée sur la caryocinèse végétale par les hautes pressions hydrostatiques et par les agents mitoclasiques. Coll. Int. CNRS. L'action mitotique et caryoclasique de substances chimiques, Montpellier 1959, vol. 88, pp. 51–65 (1960).

HALLEZ, P.: Embryogénie des Dendrocoeles d'eau douce (Doin, Paris 1887).

HAUSER, J.: Non-cellular regeneration processes in the integument of the Flatworm *Geoplana abundans*. Oncology *25:* 258–268 (1971).

HAY, E. D.: Dedifferentiation and metaplasia in vertebrate and invertebrate regeneration; in URSPRUNG The stability of differentiated state, pp. 85–108 (New York 1968a).

HAY, E. D.: Fine structure and origin of regeneration cells in planarians. Anat. Rec. *160:* 363 (1968b).

HIRN, M.: Sur le déterminisme de l'hétéromorphose de tête chez la Planaire marine *Cercyra hastata* O. Schmidt. Ann. Embr. Morph. *6:* 261–269 (1973).

HOARAU, F.: La régulation dans les territoires de régénération chez l'Isopode terrestre *Helleria brevicornis* Ebner. Ann. Embr. Morph. *2:* 87–103 (1969).

HOARAU, F.: Histogenèse des muscles au cours de la régénération de la patte chez *Helleria brevicornis* Ebner. Ann. Embr. Morph. *4:* 367–372 (1971).

HOARAU, F.: Comportement de l'hypoderme et progression de la différenciation au cours de la régénération d'un péréiopode chez l'Isopode terrestre *Helleria brevicornis* Ebner. Ann. Embr. Morph. *6:* 125–135 (1973).

HULL, F. M.: Regeneration and regulation of multiple heads in *Planaria maculata* and *Planaria agilis*. Anat. Rec. *72:* suppl. 86 (1938).

HYMAN, L. H.: Platyhelminthes and Rhynchocoela. The invertebrates, vol. 2 (McGraw-Hill, Maidenhead 1951).

KANATANI, H.: Formation of bipolar heads induced by demecolcine in the planarian *Dugesia gonocephala*. J. Fac. Sc. Univ. Tokyo *8:* 253–270 (1958).

KELLER, J.: Die ungeschlechtliche Fortpflanzung der Süsswasser-Turbellarien. Jen. Z. Naturw. *28:* 370–407 (1894).

KELLY, L. S.: Radiosensitivity of biochemical processes; in Fundamental aspects of radiosensitivity. Brookhaven Symposia in Biology, vol. 14, pp. 32–52 (Upton, New York 1961).

KIDO, T.: Studies on the pharynx regeneration in planarian, *Dugesia gonocephala*. I. Histological observations in translated pieces. Sc. Rep. Kanazawa. Univ. 7 (1961a).

KIDO, T.: Studies on the pharynx regeneration in planarian, *Dugesia gonocephala*. II. Histological observation in the abnormal regenerates produced experimentally. Sc. Rep. Kanazawa. Univ. 7 (1961b); cit. BRØNDSTED, H. V.: Planarian regeneration. Int. Ser. Monogr. Pure and Applied Biology. Zool. Div., vol. 42 (Pergamon Press, Oxford 1969).

KIORTSIS, V.: Potentialités du territoire patte chez le Triton. Revue suisse Zool. *60:* 301–410 (1953).

KISHIDA, Y.: Electron microscope studies on the planarian eye. II. Fine structure of the regenerating eye. Sc. Rep. Kanazawa. Univ. *12:* 111–142 (1967).

KOSCIELSKI, B.: Cytological and cytochemical investigation on the embryonic development of *Dendrocoelum lacteum*. Zoologica Poloniae. *16:* 83–96 (1966).

KRATOCHWIL, K. W.: Die Einwirkung von Röntgenstrahlen auf die Differenzierung in der Regeneration von *Euplanaria gonocephala*. Z. wiss. Zool. *167:* 215–237 (1962).

KRITCHINSKAYA, E. B. et LENICQUE, P. M.: Distribution des néoblastes le long du corps de la planaire *Dugesia tigrina* et pouvoir de développement de différents morceaux isolés de celui-ci. Acta zool. *50:* 69–76 (1969).

KRITCHINSKAYA, E. B. and MALIKOVA, I. G.: Reparation processes in asexual reproduction and regeneration in planaria *(Dugesia tigrina)*. Ark. Anat. Histol. Embryol. SSSR *57:* 48–51 (1969).

KÜKENTHAL, W.: Plathyhelminthes; in Handbuch der Zoologie, vol. 2 (de Gruyter, Berlin 1928–1933).

LAMEERE, A.: Précis de zoologie, vol 2, 2ème ed. 1932 (Desoer, Liège).

LANG, P.: Über die Regeneration bei Planarien. Arch. Mikr. Anat. *79:* 361–426 (1912).

LANGE, C. S.: Observation on some tumors found in two species of planaria: *Dugesia etrusca* and *D. ilvana*. J. Embryol. exp. Morph. *15:* 125–130 (1966).

LASH, J. W.: Tissue interaction and specific metabolic responses. Chondrogenetic induction and differentiation; in LOCKE Cytodifferentiation and macromolecular synthesis, pp. 235–260 (Academic Press, New York 1963).

LAZARD, L.: Restauration de la régénération de membres irradiés d'Axolotl par des greffes hétérotypiques d'organes divers. J. Embryol. exp. Morph. *18:* 321–342 (1967).

LE MOIGNE, A.: Etude du développement embryonnaire de *Polycelis nigra*. Bull. Soc. zool. Fr. *88:* 403–421 (1963).

LE MOIGNE, A.: Etude du développement et de la régénération embryonnaires de *Polycelis nigra* et *Polycelis tenuis* Turbellariés Triclades. Ann. Embr. Morph. *2:* 51–70 (1969).

LE MOIGNE, A. et GABRIEL, A.: Action de l'actinomycine D sur la différenciation cellulaire au cours de la régénération de Planaires qui viennent d'éclore. II. Etudes autoradiographiques histologiques et ultrastructurales de l'action de l'antibiotique sur les synthèses d'ARN. Z. Zellforsch. mikrosk. Anat. *115:* 442–460 (1971).

LE MOIGNE, A.; SAUZIN, M. J.; LENDER, T. et DELAVAULT, R.: Quelques aspects des ultrastructures du blastème de régénération et des tissus voisins chez *Dugesia gonocephala* (Turbellarié, Triclade). C. r. Séanc. Soc. Biol. *159:* 530–538 (1965).

LENDER, T.: Le rôle inducteur du cerveau dans la régénération des yeux d'une Planaire d'eau douce. Bull. biol. Fr. Belg. *86:* 140–215 (1952).

LENDER, T.: Sur l'inhibition de la régénération du cerveau de la Planaire *Polycelis nigra*. C. r. hebd. Séanc. Acad. Sci., Paris *241:* 1863–1865 (1955).

LENDER, T.: L'inhibition de la régénération du cerveau des Planaires *Polycelis nigra* et *Dugesia lugubris* en présence de broyats de têtes ou de queues. Bull. Soc. zool. Fr. *81:* 192–199 (1956).

LENDER, T.: L'inhibition spécifique de la différenciation du cerveau des Planaires d'eau douce en régénération. J. Embryol. exp. Morph. *8:* 291–301 (1960).

LENDER, T. et GABRIEL, A.: Etude histochimique des néoblastes de *D. lugubris* (Turbellarié, Triclade) avant et pendant la régénération. Bull. Soc. zool. Fr. *85:* 100–110 (1960).

LENDER, T. et GABRIEL, A.: Le comportement des néoblastes pendant la régénération de la Planaire *Dugesia lugubris.* Bull. Soc. zool. Fr. *86:* 67–69 (1961).

LENDER, T. et GABRIEL, A.: Les néoblastes marqués par l'uridine tritiée migrent et édifient le blastème de régénération des Planaires d'eau douce. C. r. hebd. Séanc. Acad. Sci., Paris *260:* 4095–4097 (1965).

LINDH, N. O.: The mitotic activity during early regeneration in *Euplanaria polychroa.* Ark. Zool. *10:* 497–509 (1957).

LINDH, N. O.: Histological aspects on regeneration in *Euplanaria polychroa.* Ark. Zool. *11:* 89–103 (1958).

LOEB, J.: Untersuchungen zur physiologischen Morphologie der Tiere, vol. 1 (Würzburg 1891).

LOUTIT, J. F.: Immunological and trophic functions of lymphocytes. Lancet *ii:* 1106–1108 (1962).

LUS, J.: Studies of regeneration and transplantation in the Turbellaria. I. Polarity and heteromorphoses in fresh-water Planaria. Bull. mos. obshch. ispyt. prirody, otdel biol. *32* (1924).

LUS, J.: Regenerationsversuche an marinen Tricladen. Roux Arch. EntwMech. Org. *108:* 203–227 (1926).

McLOUGHLIN, C. B.: Mesenchymal influences on epithelial differentiation. Symp. Soc. exp. Biol. *17:* 341–357 (1963).

McWHINNIE, M. A.: The effects of colchicine on reconstitutional development in *Dugesia dorotocephala.* Biol. Bull. *108:* 54–65 (1955).

McWHINNIE, M. A. and GLEASON, M. M.: Histological changes in regenerating pieces of *Dugesia dorotocephala* treated with colchicine. Biol. Bull. *112:* 371–376 (1957).

MANDEL, P. et RODESCH, J.: Aspects biochimiques d'une irradiation totale. Radiobiologie appliquée, vol. 1, pp. 359–409 (Gauthier-Villars, Paris 1966).

MANELLI, H. e NEGRI, A.: Colture *in vitro* di blastemi regenerativi di *Planaria torva.* Boll. Zool. *29:* 788–803 (1962).

MARTELLI, M.: Action de substances injectées dans le parenchyme de planaires d'eau douce. Ann. Faculté Sci. Marseille *43:* 81–87 (1970).

MARTELLI, M. et CHANDÉBOIS, R.: The functions of the type I cell system and the problems of oncogenesis in Planarians. Oncology *28:* 274–288 (1973).

MATTIESEN, E.: Ein Beitrag zur Embryologie der Süsswasser-Dendrocoelen. Roux Arch. EntwMech. Org. *38:* 331–354 (1883).

MEDAWAR, P. B.: Transplantation immunity and subcellular particles. Ann. N. Y. Acad. Sci. *68:* 255–267 (1957).

METCALF, D.: Functional interactions between the thymus and the other organs; in METCALF and DEFENDI The thymus, vol. 2, pp. 53–72 (Wistar. Inst. Press. Wist. Inst. Monogr. 1964).

MILLER, J. F. A. P. and OSOBA, D.: Current concepts of the immunological function of the thymus. Physiol. Rev. *47:* 437–520 (1967).

MONROY, A.: La rigenerazione bipolare in segmenti di arti isolati di *Triton cristatus.* Arch. ital. Anat. Embriol. *68:* 123 (1942).

MORGAN, T. H.: The control of heteromorphosis in *Planaria maculata*. Roux Arch. EntwMech. Org. *17:* 693–695 (1904).

MORITA, M.; BEST, J. B., and NOEL, J.: Microscopic studies of Planarian regeneration. I. Fine structure of neoblasts in *Dugesia dorotocephala*. J. Ultrastruct. Res. *27:* 7–23 (1969).

MURRAY, M. R.: The cultivation of Planarian tissues *in vitro*. J. exp. Zool. *47:* 467–498 (1927).

MURRAY, M. R.: The calcium-potassium ratio in culture media for *Planaria dorotocephala*. Physiol. Zool. *1:* 137–146 (1928).

MURRAY, M. R.: *In vitro* studies of planarian parenchyma. Arch. exp. Zellforsch. *11:* 656–668 (1931).

NEEDHAM, A. E.: The growth process in animals (Pitman, London 1964).

NEEDHAM, A. E.: Regeneration in the Arthropoda and its endocrine control; in KIORTSIS and TRAMPUSCH Regeneration in animals, pp. 283–323 (North-Holland, Amsterdam 1965).

NENTWIG, M. R. and SCHAUBLE, M. K.: Influence of the nutritional state on repeated head regeneration, growth and fission in the Planarian, *Dugesia dorotocephala*. J. exp. Zool. *187:* 295–302 (1974).

NEWTH, D. R.: On regeneration after the amputation of abnormal structures. I. Defective amphibian tails. J. Embryol. exp. Morph. *6:* 297–307 (1958).

OKADA, Y. K. and SUGINO, H.: Transplantation experiments in *Planaria gonocephala*. Proc. Imp. Acad. *10:* 37–40; 107–110 (1934).

OKADA, Y. K. and SUGINO, H.: Transplantation experiments in *Planaria gonocephala* Dugès. Jap. J. Zool. *7:* 373–439 (1937).

OSBORNE, P. J. and MILLER, A. T.: Acid and alkaline phosphatase changes associated with feeding, starvation and regeneration in Planarians. Biol. Bull. *124:* 285–292 (1963).

PEDERSEN, K. J.: Cytological studies on the planarian neoblast. Z. Zellforsch., Abt. Histochem. *50:* 799–817 (1959).

PEDERSEN, K. J.: Studies on the nature of planarian connective tissue. Z. Zellforsch., Abt. Histochem. *53:* 569–608 (1961).

PEDERSEN, K. J.: Studies on regeneration blastemas of the planarian *Dugesia tigrina* with special reference to differentiation of the muscle-connective tissue filament system. Roux Arch. EntwMech. Org. *169:* 134–169 (1972).

PRENANT, M.: Recherches sur le parenchyme des Plathelminthes. Arch. Morph. gén. exp. *5:* 1–175 (1922).

RAND, H. W. and BOYDEN, A. E.: Inequality of the two eyes in regenerating planarians. Zool. Jb. *36:* 68–80 (1913).

RANDOLPH, H.: The regeneration of the tail of *Lumbriculus*. J. Morph. *7:* 317–344 (1892).

REISINGER, E.: Anormogenetische und parasitogene Syncytiumbildung bei Turbellarien. Protoplasma, Wien *50:* 627–643 (1959).

RODRIGUEZ, L. V. and FLICKINGER, R. A.: Bipolar head regeneration in planaria induced by chick embryo extracts. Biol. Bull. *140:* 117–124 (1971).

ROMIEU, M. et STAHL, A.: Les problèmes histologiques et physiologiques posés par l'étude du thymus. Biol. méd. *6:* 147–155 (1949).

ROSE, C. and SHOSTAK, S.: The transformation of gastrodermal cells to neoblasts in regenerating *Phagocata gracilis*. Expl. Cell Res. *50:* 553–561 (1968).

ROSE, S. M.: Regeneration: key to understanding normal and abnormal growth and development (Appleton Century Crofts, New York 1970).

RUSTIA, C. P.: The control of biaxial development in the reconstitution of *Planaria*. J. exp. Zool. *42:* 111–142 (1924).

SANDOZ, H.: Sur la régénération antérieure chez le Némertien *Tetrastemma vittatum*. C. r. hebd. Séanc. Acad. Sci., Paris *260:* 4091–4092 (1965).

SANTOS, F. V.: Studies on transplantation in planaria. Biol. Bull. *57:* 188–197 (1929).

SANTOS, F. V.: Studies on transplantation in planarians. Physiol. Zool. *4:* 111–164 (1931).

SAUZIN, M. J.: Etude au microscope électronique du néoblaste de la planaire *Dugesia gonocephala* (Turbellarié, Triclade) et de ses changements ultrastructuraux au cours des premiers stades de la régénération. C. r. hebd. Séanc. Acad. Sci., Paris *263:* 627–629 (1966).

SAUZIN, M. J.: Etude ultrastructurale de la différenciation du néoblaste au cours de la régénération de la Planaire *Dugesia gonocephala*. I. Différenciation en cellule nerveuse. Bull. Soc. zool. Fr. *92:* 313–318 (1967a).

SAUZIN, M. J.: Etude ultrastructurale de la différenciation au cours de la régénération de la Planaire *Dugesia gonocephala*. II. Différenciation musculaire. Bull. Soc. zool. Fr. *92:* 613–616 (1967b).

SAUZIN-MONNOT, M. J.: Etude ultrastructurale des néoblastes de *Dendrocoelum lacteum* au cours de la régénération. J. Ultrastruct. Res. *45:* 20–26 (1973).

SAYLES, L. P.: Double nucleoli and mitosis in cells of the alimentary tract of *Lumbriculus* following dilution of the body fluids. J. exp. Zool. *58:* 487–494 (1931).

SCHEWTSCHENKO, N. N.: Die Wechselwirkung von Teilen von verschiedener physiologischer Aktivität bei Planarien. Biol. Zbl. *6:* 581–587 (1937).

SCHILT, J.: Greffes hétéropolaires chez les Planaires. Région pharyngienne. C. r. Séanc. Soc. Biol. *162:* 2263–2265 (1968).

SCHILT, J.: Induction expérimentale d'excroissances par des greffes hétéropolaires chez la Planaire *Dugesia lugubris*. Ann. Embr. Morph. *3:* 93–106 (1970).

SCHULTZ, E.: Über Reduktionen. I. Über Hungererscheinungen bei *Planaria lactea*. Roux Arch. EntwMech. Org. *18:* 555–577 (1904).

SEILERN-ASPANG, F.: Beobachtungen über Zellwanderungen bei Tricladengewebe in der Gewebekultur. Roux Arch. EntwMech. Org. *152:* 35–42 (1960a).

SEILERN-ASPANG, F.: Syncytiale und differenzierte Tumoren bei Tricladen. Roux Arch. EntwMech. Org. *152:* 517–523 (1960b).

SEILERN-ASPANG, F. and KRATOCHWIL, K. W.: Relation between regeneration and tumor growth; in KIORTSIS and TRAMPUSCH Regeneration in animals, pp. 452–473 (North-Holland, Amsterdam 1965).

SENGEL, C.: Culture *in vitro* de blastème de régénération de Planaires. J. Embryol. exp. Morph. *8:* 468–476 (1960).

SENGEL, C.: Culture *in vitro* de blastèmes de régénération de la planaire *Dugesia lugubris*. Ann. Epiphyties *14:* 173–183 (1963).

SENGEL, P.: Sur les conditions de la régénération normale du pharynx chez la Planaire d'eau douce *Dugesia lugubris*. Bull. biol. Fr. Belg. *85:* 376–391 (1951).

SENGEL, P.: Sur l'induction d'une zone pharyngienne chez la Planaire d'eau douce *Dugesia lugubris*. Arch. Anat. micr. Morph. exp. *42:* 57–66 (1953).

SILBER, H. and HAMBURGER, V.: The production of *duplicitas cruciata* and multiple heads by regeneration in *Euplanaria tigrina*. Physiol. Zool. *12:* 285–300 (1939).

SIVICKIS, F. B.: A quantitative study of regeneration in *Dendrocoelum lacteum*. Ungar. biol. Forschungsinst. *4:* 1 (1931).

SKAER, R. J.: Some aspects of the cytology of *Polycelis nigra*. Q. J. micr. Sci. *5:* 295–318 (1961).

SKAER, R. J.: The origin and continous replacement of epidermal cells in the planarian *Polycelis tenuis* (Iijima). J. Embryol. exp. Morph. *13:* 129–139 (1965).

SPERRY, P. J. and ANSEVIN, K. D.: Determination in regenerating tissues of *Dugesia dorotocephala:* the influence of nerve cord grafts. J. Embryol. exp. Morph. *33:* 85–93 (1975).

SPERRY, P. J.; ANSEVIN, K. D., and TITTEL, F. K.: The inductive role of the nerve cord in regeneration of isolated postpharyngeal body sections of *Dugesia dorotocephala*. J. exp. Zool. *186:* 159–174 (1973).

STEINBÖCK, O.: Regenerations- und Transplantationsversuche an *Amphiscolops* spec. (Turbellaria, Acoela). Roux Arch. EntwMech. Org. *154:* 308–353 (1963).

STEINBÖCK, O.: Regenerationsversuche mit *Hofstenia Giselae* (Turbellaria, Acoela). Roux Arch. EntwMech. Org. *158:* 394–458 (1967).

STEINMANN, P.: Prospektive Analyse von Restitutionsvorgängen. I. Die Vorgänge in den Zellengeweben und Organen während der Restitution von Planarienfragmenten. Roux Arch. EntwMech. Org. *108:* 646–679 (1926).

STÉPHAN, F.: Quelques problèmes concernant la régénération des Planaires. Bull. Acad. Soc. Lorraine Sci. *8:* 177–183 (1969).

STÉPHAN-DUBOIS, F.: Les néoblastes dans la régénération postérieure des Oligochètes Microdriles. Bull. biol. Fr. Belg. *88:* 181–247 (1954).

STÉPHAN-DUBOIS, F.: Migration et différenciation des néoblastes dans la régénération antérieure de *Lumbriculus variegatus*. C. r. Séanc. Soc. Biol. *150:* 1239–1242 (1956).

STÉPHAN-DUBOIS, F.: Le rôle des leucocytes dans la régénération caudale de *Nereis diversicolor*. Arch. Anat. micr. Morph. exp. *47:* 604–652 (1958).

STÉPHAN-DUBOIS, F.: Les néoblastes dans la régénération chez les planaires; in KIORTSIS and TRAMPUSCH Regeneration in Animals, pp. 112–130 (North Holland, Amsterdam 1965).

STÉPHAN, F. et SCHILT, J.: Etude histologique d'excroissances induites par cautérisation chez la Planaire *Dugesia tigrina*. C. r. hebd. Séanc. Acad. Sci., Paris *263:* 1732–1734 (1966).

STÉPHAN, F. et SCHILT, J.: Expériences d'autogreffes hétéropolaires chez la Planaire *Dugesia lugubris*. C. r. hebd. Séanc. Acad. Sci., Paris *264:* 3016–3019 (1967).

STEVENS, N. M.: A histological study of regeneration in *Planaria simplicissima, Pl. maculata* and *Pl. Morgani*. Roux Arch. EntwMech. Org. *24:* 350–373 (1907).

STOCUM, D. L. and DEARLOVE, G. E.: Epidermal-mesodermal interaction during morphogenesis of the limb regeneration blastema in larval salamanders. J. exp. Zool. *181:* 49–62 (1972).

STONE, L. S.: Regeneration of the retina, iris and lens; in THORNTON Regeneration in Vertebrates, pp. 3–14 (University of Chicago Press, Chicago 1958).

SUGINO, H.: Miscellany on planaria transplantation. A supplemental note to the transplantation experiments in *Planaria gonocephala*. Ann. zool. Jap. *17:* 185–193 (1938).

SUGINO, H.: Homopolar union in *Planaria gonocephala*. Jap. J. Zool. *9:* 176–183 (1941).

SUGINO, H.: Recombination experiments of small pieces in *Dugesia gonocephala*. Memoirs Osaka Univ. B. Nat. Sci. *2:* 1–44 (1953).

SUGINO, H.; OKUNO, Y.; ONO, T., and SAKUMA, E.: Studies on the regeneration of epidermis in fresh-water Planarian, *Dugesia japonica*. Memoirs Osaka Kyoiku Univ. *18:* 29–41 (1969).

SUGINO, H.; OKUNO, Y., and YOSHINOBU, J.: Effect of transplanted pieces from non-X-irradiated worms on irradiated ones in *Dugesia japonica*. Memoirs Osaka Kyoiku Univ. *19:* 63–76 (1970).

TAR, E. and TÖRÖK, L. J.: Investigations on somatic twin-formation, benignant and malignant tumors in the species *Dugesia tigrina*. Acta Biol. Acad. Sci. Hungaricae *15:* suppl. 6, p. 34 (1964).

TESHIROGI, W.: The effect of lithium chloride on head frequency in *Dugesia gonocephala*. Bull. Mar. Biol. St. Asamushi 8 (1955).

TESHIROGI, W.: Transplantation of the ganglionic region into the posterior levels of the Turbellarian *Bdellocephala brunnea*. Sci. rep. Tôhoku Univ. *22:* 197–206 (1956).

THORNTON, C. S. and THORNTON, M. T.: The regeneration of accessory parts following epidermal cap transplantation in Urodeles. Experientia *21:* 146–148 (1965).

THOUVENY, Y.: Les systèmes histogénétiques et la dédifférenciation cellulaire dans la morphogenèse des Annélides Polychètes. Arch. Zool. exp. gén. *108:* 347–520 (1967).

TRAMPUSCH, H. A. L. and HARREBOMÉE, A. E.: Dedifferentiation a prerequisite of regeneration; in KIORTSIS and TRAMPUSCH Regeneration in Animals, p. 341 (North-Holland, Amsterdam 1965).

TRAMPUSCH, H. A. L. and HARREBOMÉE, A. E.: Dedifferentiation and the interconvertibility of different cell-types in the Amphibian extremity. Acta Emb. exp. *1:* 35–39 (1969).

URBANI, E. e CECERE, F.: Incorporazione di uridina [3]-H nei neoblasti di Planaria. Rend. 1st. Sc. Camerino *5:* 106–108 (1964).

VAN CLEAVE, C. D.: The effects of X-radiation on the restitution of *Stenostomum tenuicauda* and some other worms. Biol. Bull. *47:* 304–314 (1934).

VANDEL, A.: La question de la spécifité cellulaire chez les Planaires. C. r. hebd. Séanc. Acad. Sci., Paris *172:* 1614–1617 (1921a).

VANDEL, A.: Recherches expérimentales sur les modes de reproduction des Planaires Triclades Paludicoles. Bull. biol. Fr. Belg. *55:* 343–518 (1921b).

VERHOEF, A. M.: The mitotic activity during the regeneration of *Polycelis nigra*. Proc. Kon. Ned. Ak. Wet. *49:* 548–553 (1946).

WAGNER, R. VON: Zur Kenntnis der ungeschlechtlichen Fortpflanzung von *Micro-*

stoma nebst allgemeinen Bemerkungen über Teilung und Knospung im Tier-reich. Z. Jb. *4:* 349–423 (1890).

WEIGAND, K.: Regeneration bei Planarien und *Clavelina* unter dem Einfluss von Radiumstrahlen. Z. wiss. Zool. *136:* 255–318 (1930).

WEISS, P.: Ganzregenerate aus halbem Extremitätenquerschnitt. Roux Arch. Entw-Mech. Org. *107:* 1–53 (1926).

WOLFF, E.: Les phénomènes d'induction dans la régénération des Planaires d'eau douce. Rev. suisse Zool. *60:* 540–546 (1953).

WOLFF, E.: Migrations et contacts cellulaires dans la régénération. Expl. Cell Res., suppl. 8, pp. 246–259 (1961).

WOLFF, E.: Recent researches on the regeneration of Planaria. Regeneration. 20th Growth Symposium, pp. 53–84 (Ronald, New York 1962).

WOLFF, E. et DUBOIS, F.: Sur une méthode d'irradiation localisée permettant de mettre en évidence la migration des cellules de régénération chez les Planaires. C. r. Séanc. Soc. Biol. *141:* 903–906 (1947).

WOLFF, E. et LENDER, T.: Les néoblastes et les phénomènes d'induction et d'inhibi-tion dans la régénération des Planaires. Ann. Biol. *1:* 499–529 (1962).

WOLFF, E.; LENDER, T. et ZILLER-SENGEL, C.: Le rôle des facteurs auto-inhibiteurs dans la régénération des Planaires. (Une interprétation nouvelle de la théorie des gradients physiologiques axiaux de Child.) Rev. suisse Zool. *71:* 75–98 (1964).

WOLFF, E.; SENGEL, P. et SENGEL, C.: La région caudale est-elle capable d'induire la régénération d'un pharynx? C. r. hebd. Séanc. Acad. Sci., Paris *246:* 1744–1746 (1958).

WOLFF, E. et WEY-SCHUÉ, M.: Démonstration expérimentale de la migration des cel-lules de régénération des membres de *Triton cristatus*. C. r. Séanc. Soc. Biol. *146:* 113–117 (1952).

WOLPERT, L.: Positional information and the spatial pattern of cellular differentia-tion. J. theor. Biol. *25:* 1–47 (1969).

WOLPERT, L.: Positional information and pattern formation. Curr. Topics Develop. Biol. *6:* 183–224 (1971).

WOLSKY, A.: Starvation and regeneration potency in *Dendrocoelum lacteum*. Nature, Lond. *135:* 102 (1935).

WOLSKY, A.: The effect of chemicals with gene-inhibiting activity on regeneration. In SHERBET Neoplasia and cell differentiation, pp. 153–188 (Karger, Basel 1974).

WOODRUFF, L. and BURNETT, A. L.: The origin of the blastema cells in *Dugesia ti-grina*. Expl. Cell Res. *38:* 295–305 (1965).

YAMADA, T.: Cellular synthetic activities in induction of tissue transformation; in Cell differentiation. Ciba Fdn Symp., pp. 116–130 (Little, Brown, Boston 1967).

YNTEMA, C. L.: Blastema formation in sparsely innervated and aneurogenic forelimbs of *Amblystoma* larvae. J. exp. Zool. *142:* 423–440 (1959).

ZILLER, C.: La régénération du pharynx chez la planaire *Dugesia tigrina*. C. r. hebd. Séanc. Acad. Sci., Paris *277:* 1365–1368 (1973).

ZILLER, C.: Activation de la régénération chez les Planaires par des amputations préalables. Aptitude des planaires activées à régénérer en présence d'actino-mycine D. C. r. hebd. Séanc. Acad. Sci., Paris *278:* 2347–2350 (1974a).

ZILLER, C.: Activation de la régénération chez les planaires par des amputations préalables. Etendue et spécificité de l'activation. C. r. hebd. Séanc. Acad. Sci., Paris *278:* 2971–2974 (1974b).

ZILLER-SENGEL, C.: Inhibition de la régénération du pharynx chez les Planaires; in KIORTSIS and TRAMPUSCH Regeneration in Animals, pp. 193–201 (North-Holland, Amsterdam 1965).

ZILLER-SENGEL, C.: Sur le facteur inhibiteur de la régénération du pharynx chez les Planaires d'eau douce. Bull. Soc. Zool. Fr. *92:* 295–302 (1967a).

ZILLER-SENGEL, C.: Recherches sur l'inhibition de la régénération du pharynx chez les Planaires. I. Mise en évidence d'un facteur auto-inhibiteur de la régénération du pharynx. J. Embryol. exp. Morph. *18:* 91–105 (1967b).

ZILLER-SENGEL, C.: Recherches sur l'inhibition de la régénération du pharynx chez les Planaires. II. Variations d'intensité du facteur inhibiteur suivant les espèces et les phases de la régénération. J. Embryol. exp. Morph. *18:* 107–117 (1967c).